Practical Manual of
WASTEWATER CHEMISTRY

Barbara A. Hauser

CRC Press
Taylor & Francis Group
Boca Raton London New York

CRC Press is an imprint of the
Taylor & Francis Group, an informa business

Published 1996 by CRC Press
Taylor & Francis Group
6000 Broken Sound Parkway NW, Suite 300
Boca Raton, FL 33487-2742

© 1996 by Taylor & Francis Group, LLC
CRC Press is an imprint of Taylor & Francis Group, an Informa business

First issued in paperback 2019

No claim to original U.S. Government works

ISBN 13: 978-0-367-44854-7 (pbk)
ISBN 13: 978-1-57504-012-7 (hbk)

**Visit the Taylor & Francis Web site at
http://www.taylorandfrancis.com**

**and the CRC Press Web site at
http://www.crcpress.com**

Library of Congress Cataloging-in-Publication Data

Hauser, Barbara A.
    Practical manual of wastewater chemistry / Barbara A. Hauser.
      p. cm.
    Includes bibliographical references and index.
    ISBN 1-57504-012-3
    1. Sewage—Analysis—Handbooks, manuals, etc. I. Title

TD735.H38 1996
628.3—dc20

96-20309
CIP

Library of Congress Card Number 96-20309

Table of Contents

Introduction

A colorless, odorless, tasteless liquid, water is the only common substance that occurs naturally on the earth in all three physical states: solid, liquid, and gas. Seventy three percent of the earth's surface is covered with it, almost 328,000,000 cubic miles. The human body is seventy percent water by weight; it is essential to the life of every living thing.

This is a unique molecule; one side of it is positive, the other is negative. Water molecules orient to neutralize electric charges; this buffering action keeps foreign ions from reacting and precipitating. Pure water itself resists ionizing, and is a poor electrical conductor. These properties give water an amazing capacity to dissolve other molecules; it is in fact considered the universal solvent.

Man has always used water to carry away his wastes, and in past centuries nature has been able to convert those wastes back to life's usable compounds for future generations. Now man must aid in that conversion; wastewater must be treated to a quality that will allow natural systems to complete the recycling so necessary to our existence.

This manual was prepared to enhance the wastewater treatment plant operator's understanding of laboratory theory and procedure in the testing of wastewater, to appreciate the nature of the contaminants under analysis, the analytical methods, and their limitations. When interpreting lab data for process control, test results are a reliable description of the sample analyzed; they define the status of the treatment process. Treatment can be optimized, toxic inputs and process upsets identified. State and federal regulatory compliance will be achieved.

Methods in this guide are EPA approved (most often from Standard Methods). Included is an explanation of each analysis and the equipment used to perform it, to aid in understanding the chemistry of the reactions, and the working mechanism of the instruments. A summary of treatment plant significance for each analysis is included, along with a chart of troubleshooting suggestions. At the back of the guide are typical math exercises, arranged by chapter.

Sampling and Laboratory Safety

Sampling

All laboratory testing depends first upon proper sampling technique. Sampling must be designed to obtain accurate data for identifying treatment process changes, and influent and effluent water qualities. Representative sampling is important. The objective is to remove a small portion which is representative of the entire flow, and adequately reflects actual conditions in the water.

Sample Location

The test requirements usually determine sample location (ex. raw, primary effluent, final). Take the sample where mixing is best, and the water is of uniform quality. Sampling location must be accessible; safe entry to confined spaces is imperative. Avoid slippery surfaces; do not climb on or under guardrails.

Composite Samples

This type of sample is taken to determine average conditions in a large volume of water whose chemical properties vary significantly over the course of a day. These are usually flow proportioned, and collected with an automatic sampler. Small aliquots are taken at regular intervals and pooled into one large sample over a 24 hour period. If composite samples are to be taken manually, the frequency of sampling should depend on the number of changes in water quality. Most often, grabs are taken hourly, and the volume of each is flow proportioned; samples are then combined into a composite for testing.

Grab Samples

A grab sample is taken all at once, at a specific time and place. In wastewater treatment grabs are usually taken at peak flow conditions. Bacteriologic samples, temperature and dissolved oxygen samples are always grabs. Insert container upside down into the water. Rotate open end toward direction of flow, and allow to fill under the surface. Sample about 12 inches underwater, or at center of the channel, about medium depth from the bottom. Avoid surface scum and bottom sediment. Be careful not to collect deposits from tank sidewalls. If large, uncharacteristic particles enter the sample, eliminate them. Fill sample bottle completely to exclude air space if sample is to be analyzed for DO, NH_3, H_2S, SO_2, pH, Chlorine Residual, Alkalinity or VOC's. If preservative or dechlorinating agent has been added to empty sample bottle, adjust sampling technique so that the bottle does not overfill, or the chemical will be washed out. Be careful to handle sterile bottles for bacteriologic testing aseptically. Collect enough sample to allow duplicate and spiked analysis. Sludge samples are heaviest at the beginning of pumping, or from the bottom-most tap. Try for a representative portion of the flow.

If an influent and an effluent sample are desired from a process unit with a detention time of two hours, consider the timing. Take the influent sample now, and take the effluent sample in two hours. There will be a good chance of getting the same water.

The Sample Bottle

For chemical testing, the sample bottle must be clean. For bacteriologic testing, the sample bottle must be clean, and sterile. Bottles may be glass or plastic for most analyses; labels must be firmly attached to the sample bottle, not to the lid. Use labels that will not come off when damp. Use water insoluble ink pen. The label on a sample bottle should include sample ID number, date and time of collection, type of sample (raw, final), location (east aerator), adverse weather conditions, collector's initials, analysis to be performed, sample preservation, if any.

Preservation and Transport

Dissolved oxygen, pH, temperature should be analyzed onsite, at the sampling location. All samples should be analyzed as soon as possible after collection.

Chain of Custody

This is a legal requirement, and refers to the recorded handling of a sample from collection to analysis. It allows for tracing individual samples in the case of a problem. The sample is in custody if it is in hand or in sight, if it is locked away or placed in a secure area where nobody can enter without the possessor's knowledge. When the sample changes hands, a Change of Possession form is signed by both parties; date and time are recorded.

The Chain of Custody Records include:
- Sample Labels—for sample identification
- Sample Seals—for shipped samples, to ensure no tampering.
- Field Logbook—includes all information on label, container type, sample size, field analysis, number of samples taken.
- Chain of Custody Record—includes label information and Change of Possession forms.

Preservation Techniques and Holding Times

Test	Preservation	Max. Holding Time
Coliform (Total & Fecal)	4 deg.C, .008% $Na_2S_2O_3$	6 hrs.
Acidity	4 deg.C	14 days
Alkalinity	---	---
Ammonia	4 deg.C, H_2SO_4 to pH2	28 days
BOD	4 deg.C	48 hrs.
COD	4 deg.C, H_2SO_4 to pH2	28 days
Chloride	---	---
Chlorine Residual	---	do immed.
Color	4 deg.C	48 hrs.
Fluoride	---	28 days
Hardness	HNO_3 to pH2	6 months
pH	---	do immed.
Metals	HNO_3 to pH2	6 months
Mercury	HNO_3 to pH2	28 days
Nitrate	4 deg.C	48 hrs.
Nitrite	4 deg.C	48 hrs.
Orthophosphate	4 deg.C, filter	48 hrs.
Total Phosphorus	4 deg.C, H_2SO_4 to pH2	28 days
DO	---	do immed.
Total Solids	4 deg.C	7 days
Dissolved Solids	---	48 hrs.
Total Suspended Solids	---	7 days

continued

Test	Preservation	Max. Holding Time
Settleable Solids	---	48 hrs.
Specific Conductance	---	28 days
Sulfate	---	---
Temperature	---	do immed.
Turbidity	4 deg.C	48 hrs.

Online Sampling and Analysis

There is increasing use of automatic online sampling and monitoring. It has the advantage of yielding constant and immediate results. Time is saved, and problems due to carriage to laboratory are eliminated.

Meters are installed directly into the process flow, and the water is monitored as it flows by. Typical tests that are performed this way are pH, DO, temperature, conductivity. The signal may be read onsite, or transmitted to the lab or control room for remote readout. Frequent cleaning and calibration of the monitoring instruments are most important.

Laboratory Records

Process control and regulatory compliance depend upon the proper recording of laboratory analysis data.

Analysis Reports (Bench Sheets) should include:
• Name, Time and Date of analysis
• Analyst name
• Sample preparation
• Analysis method
• Test conditions (stds, reagents, instrument settings, temperature, reaction time)
• Results of Analysis
• Observations - comments

Laboratory Records must be kept for 5 years.

Laboratory Safety

Safety is just as important in the laboratory as it is in the treatment plant. A number of hazards exist there; be alert and careful, and aware of potential dangers at all times. Although very dilute solutions are being analyzed, the laboratory chemist does handle some extremely strong acids and bases, and some toxics such as arsenic, chromium, mercury. It is vital to have a thorough knowledge of safe operating procedures for each analysis.

Twenty-two Lab Safety Rules

1. Know where the safety equipment is: safety shower, eye wash, fire extinguisher, telephone, first aid kit, acid spill kit. In emergencies response must be automatic, for it may be difficult to think clearly.
2. No eating, drinking, smoking in the lab.
3. Do not attempt to taste or smell any chemical.
4. Use pipet bulbs for pipetting.
5. Never handle dry chemicals with the hands. Use a spatula.
6. Never stopper a flask with the thumb to mix the contents. This is horrible laboratory technique; it will contaminate the test, and chemical in contact with the skin or mouth could be dangerous.
7. Wash hands well with soap and water before leaving the lab. There are chemical toxins and biological pathogens there.
8. Wear Safety glasses at all times in the lab. Contact lenses should not be worn in the lab.
9. Label all prepared solutions properly: chemical name, concentration, date prepared, chemist's initials. If a solution is over 1% concentration (.1% if it is a carcinogen), OSHA requires that labels also list fire and health hazard (special labels may be purchased for this purpose). Most liquids in a wastewater laboratory look like water; if the concentration is not on it, it is useless as a chemical, and it could be dangerous.
10. Add Acid To Water! When diluting concentrated acids, always put the water in the beaker first, then add the acid, SLOWLY. Acids are hydroscopic; they react quickly with the water they are dissolving in, creating heat. When there is a large quantity of acid mixing with a little water, the reaction can be violent.
11. Strong Oxidizing Agents such as Ammonium Persulfate and Potassium Dichromate can produce violent reactions. Use carefully! Store separately.

12. Oil, grease, mercury, volatile solvents and strong acids should be kept out of sinks. Chemical incompatibility may cause an explosion from trapped vapors in the drains.

13. Be Aware of Heat! Don't touch a hot plate to see if it is on; assume that it is. Never leave a heating solution. Dissolve strong acids and bases slowly. They create great heat, and may spatter, or may break the glassware. When opening a furnace, oven, or hot water bath, stand away; use tongs; wear gloves.

14. Strong acids and bases emit choking fumes. Use them under the fume hood. Do not lean over a boiling solution. It may be emitting toxic vapors. Boil solutions under the fume hood. If fumes are unexpectedly encountered, get to the nearest source of fresh air.

15. Acids burn! Strong acids and bases are highly corrosive, especially to the skin. Handle with extreme care to avoid contact. Wash off with plenty of running water. Use the safety shower. If in the eyes, use the eye wash. Call for help. Seek medical attention.

16. Hold and carry all large chemical containers with two hands to minimize risk of dropping; hands and/or glassware are often wet.

17. Don't try to pick up broken glass with the fingers. Sweep it up with a broom. Dispose of in a special container labeled "broken glass".

18. Some Chemicals Are Toxic!
Arsenic: highly toxic; avoid inhalation, ingestion and skin exposure. Prepare in a fume hood.
Azides: sodium azide (NaN_3) is toxic, and reacts with acid to produce the more toxic hydrazoic acid. Avoid inhalation, ingestion, and skin exposure.
Cyanides: most are toxic; avoid ingestion. Do not acidify cyanide solutions; toxic HCN gas is produced.
Mercury: liquid mercury is a highly toxic and volatile element; prevent inhalation and skin exposure. Mercury compounds are also toxic.
Organics: many are toxic and/or carcinogenic; many are flammable or explosive.

19. Beware of electrical hazards. Do not plug in electrical equipment with wet hands. When turning on a hot plate, be sure that no electrical cords are touching the heater plate. The cord will melt and cause an electrical short.

20. Wastewater Contains Pathogens. When handling wastewater, especially with a hand cut or abrasion, use medical laboratory gloves. Anyone working in a wastewater treatment plant should keep immunizations current.

21. Compressed gas is not a toy. Keep cylinders chained to the wall. Do not fool with pressure regulators. When using, open cylinder valves slowly.
22. Do not try to change the lab test procedure. Chemicals added, subtracted, or mixed in a different order may cause explosive conditions.

Good Housekeeping is Part of Safety. Be scrupulously clean with the glassware and chemicals; contaminated glassware will ruin the lab test, and it can be hazardous.

Quality Assurance/Quality Control

A written and practiced Quality Assurance/Quality Control program is now required at both water and wastewater treatment plants in the United States, stated in the operating permits which these plants hold. Methods may vary somewhat from plant to plant, but the goal is to control and assure the quality of laboratory analyses.

When dealing with water samples, there is no written "right answer" which states how much phosphorus, or chlorine residual, or nitrate is in the sample. Chemical test results are used to control the treatment process, and these results are sent to the state for compliance reporting. The treatment plant laboratory cannot afford to be wrong. In a quality control program, definite statistical procedures and calculations are used to provide both precision and accuracy to laboratory work. This makes the data reliable.

Precision: Take a sample of wastewater treatment plant effluent, break it into 50 equal aliquots, and test all 50 for ammonia. If exactly the same result is obtained on all of them, that's precision.

Accuracy: When doing the above test, if the right result was also obtained with each of the 50 tests, that's accuracy. In our situation of testing water samples, strict adherence to standard methods, along with troubleshooting possible analysis and sampling problems, will provide reliability and the "right" result.

The goal is to have the right result, and to have it all the time.

Methods

Replicate Analyses—for *precision*

Run more than one sample of the same water through the entire test procedure. Differing results may show variation within a test method.

This may apply to sampling or handling procedure, to changes within the sample, or to the instrument used. It will then take some troubleshooting to determine the cause.

Split Samples—for *accuracy*

Split a sample in two and have one operator run one half, another operator run the other half. Alternately, share half with a neighboring plant. Either way, the result of both halves should be the same.

Reference Samples—for *accuracy*

Send to EPA or to the state lab for a sample which has been previously tested, and has a known amount of the constituent in it. Then test for that constituent; it should have the same result (sometimes these are referred to as reference standards).

Standard Samples—for *accuracy*

Make up a standard with a known amount of the constituent in it. Run this standard along with today's sample. If the test result from the standard is exactly the concentration that it was made up at, then today's sample result is most likely correct also.

Spiked Samples—for *accuracy*

Add a known amount of the constituent to be measured for to your sample. Run one sample with the spike in it, and one sample without the spike. For example, in testing sample for phosphorus - add 1ppm phosphorus to one sample; the other sample is run with no addition. If the unspiked sample result is .4 ppm phosphorus, then the spiked sample result has to be 1.4 ppm phosphorus. If the spiked sample result is lower, then the test is only showing a percentage of the natural phosphorus concentration, and of the spiked phosphorus concentration. (the same applies if the result comes out higher).

Differences may result from chemical interferences in the sample, which would not show up by simply running a standard along with the sample, as in the previous method.

Armed with QA/QC data, one can then establish allowable deviations.

Percent Recovery: This statistical calculation is called *Percent Recovery* of *Standard.* Using spiked samples, it can be determined what percent of the standard is recovered as a test result. 100% is complete recovery. *90% - 110% is considered acceptable.* Outside this range, errors should be corrected.

For example, samples have been tested for phosphorus on ten consecutive days. Each day an unspiked sample was run, along with a sample spiked with 1 ppm phosphorus.

Plain Sample (ppm)	Spiked Sample (ppm)	% Recovery
1.02	1.93	91%
.23	1.21	
.03	.96	
.64	1.61	
2.01	3.01	
.07	1.06	
.03	1.01	
2.01	2.96	
1.01	1.93	
.04	1.05	

$$\% \text{ Recovery} = \frac{\text{Spiked} - \text{Sample}}{\text{Amount Spiked (ppm)}}$$

Calculate the Percent Recoveries. How often do the results fall out of acceptable range? Average the Percent Recoveries for a month. Is the average acceptable? Determine where the results stand, statistically. This procedure checks accuracy of results. Done a number of times over, it also checks precision. Percent Recovery can be done on replicate samples, as well as on different samples.

These statistical methods only help to determine whether and by how much something is wrong. They do not offer a solution to the problem. However, recognition of the error is the first step. Continue to use the same statistical method to check accuracy as it improves, until all results are in an acceptable range.

Standard Deviation

Standard Deviation is a statistical calculation which is done to check precision; it is defined as "variance from the mean". For example, take a group of 20 replicate samples tested for BOD. All the results should have come out the same. If they don't, it must first be determined which is most likely the correct result. On a group of replicates like this, the best that can be done is to assume that the average of the results is most likely closest to the true answer. The larger the group of replicates done, the more assurance there is of this. (This is assuming that accuracy of procedure is correct, and only precision is off).

The Standard Deviation is the amount by which any one result varies from the average. The calculation assigns a number value to that amount of variance which is called One Standard Deviation (1S). If the sample being considered varies from the mean by no more than 1S (+ or -), then that sample result is acceptable. If the sample being considered varies from the mean by as much as 2S, steps should be taken to try to determine the cause of the sample variance. This is the "warning limit". If the sample being considered varies from the mean by as much as 3S, then this is bad data. 3S is the control limit.

Calculating Standard Deviation:

$$S = \sqrt{\frac{(\Sigma x^2) - \frac{(\Sigma x)^2}{n}}{n-1}}$$

Each Result: x Sum of Results Squared: $(\Sigma x)^2$
Number of Results: n Each Result Squared: x^2
Sum of Results: Σx Sum of Squared Results: Σx^2

Data:

20 ppm	21 ppm	20 ppm	18 ppm
18 ppm	19 ppm	27 ppm	19 ppm
20 ppm	18 ppm	20 ppm	21 ppm
20 ppm	21 ppm	21 ppm	21 ppm
19 ppm	20 ppm	23 ppm	17 ppm

$n = 20$ $\Sigma x = 401$
$(\Sigma x)^2 = 160801$ $\Sigma x^2 = 8139$
$\bar{x} = 20$ 1S = 2.3

A *Control Chart*, a graphical representation of the data and variances from the mean is constructed, Enter 1S, 2S, 3S up and down the left side (+ and − from the mean). Enter data and construct curve.

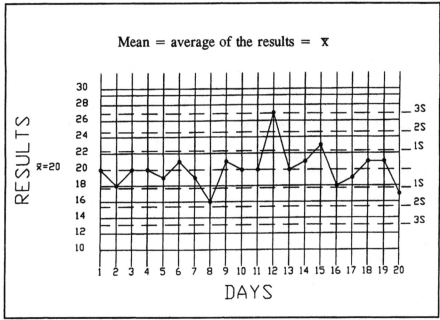

Figure 1 - Control Chart

Reasons for Inaccuracies: Errors

There is inherent human error in everything we do. The challenge is to recognize where there are errors, and then to attempt to minimize them. The quality control procedures and calculations above help to recognize the errors. The troubleshooting methods below help to solve the problems once they are recognized.

Random Error

These are not consistent, and will vary from the true result by a random number.

- Muffle furnace whose temperature controller is not working well, and temperature swings up and down.
- Measuring volumes in beakers, and using different ones each time.

- Different operators performing the same test, or taking the sample, or calibrating the meter.
- Neglecting to calibrate an instrument on a regular basis.
- Two analysts calling the endpoint of a titration at different shades of pink.
- Using outdated reagents.
- Using dirty BOD bottles.
- Neglecting to shake the sample.
- Neglecting to routinely check distilled water quality.
- Neglecting to warm up the meter before use.
- Neglecting to check temperature of incubators and ovens.
- Neglecting to standardize reagents.
- Neglecting to dry chemicals where it is called for.
- Carelessness about keeping color development time the same for all colorimetric standards and samples before reading.
- Rarely checking analytical balances with standard weights.
- Neglecting to zero the analytical balance before each use.
- Neglecting to mix reagents well enough in the volumetric flask before use.
- Making volumetric measurements with hot liquids.
- Performing a colorimetric procedure whose calibration curve is linear at 1-3 ppm, with samples and standards that range from 8-12 ppm.
- Keeping reagents for longer than their shelf life.
- Storing photo reactive chemicals on the windowsill.
- Performing a DO test on mixed liquor that has been sitting in the lab for two hours.

Systematic Error

These are consistent, and will vary from the true result by the same amount all the time.

- Using a distilled water blank instead of a reagent blank in a particular colorimetric procedure.
- Reading the meniscus on a pipet incorrectly.
- Calibrating a meter incorrectly.
- Measurement of viscous liquids and slurries in pipets and cylinders.
- Using a pipet with a chipped tip that always delivers .4ml less than its calibrated value.
- Using a ruler with the end worn off.
- Using soap containing phosphorus to wash phosphate glassware.
- Measuring 10 ml of liquid in a 100 ml graduated cylinder.

Measurements are done in parts per million, and parts per billion. Small inconsistencies can translate into large inaccuracies in analysis.

Sample WasteWater Treatment
Plant QA/QC Program

1. **BOD:**
   - set up a dilution water blank with each set of BOD's incubated.
   - run duplicate analysis on raw & final effluent samples once every two weeks.
   - run glucose/glutamic acid check once every six months.
   - calibrate meter with each use; change membrane every 2 weeks.

2. **Total Suspended Solids:**
   - run duplicate analyses of raw & final effluent every 2 weeks.
   - check balances against standard weights monthly.

3. **Fecal Coliform:**
   - run duplicate of each dilution used every other week.
   - run a blank of sterilized dilution water once per month.
   - monthly run a sample (.1ml) of unchlorinated effluent to make sure bacteria will grow.

4. **Total Phosphorus:**
   - run spiked samples weekly on raw and final; calculate % Recovery.
   - run a standard and a blank with each set of samples done; check standard reading against calibration curve. Vary the standard each week.
   - run duplicates of raw and final monthly.

5. **Overall Lab Performance:**
   - run reference samples from EPA on all tests once per year.
   - run split samples on all tests once per year.
   - check temperatures daily on muffle furnace, drying oven, BOD incubator, coliform incubator, sample refrigerator. Have at least one certified thermometer for this. Check all thermometers against it.
   - have the analytical balance serviced annually.
   - keep records of all quality assurance work done.
   - the operator doing the analyses initials the bench sheets.
   - develop and implement a lab equipment maintenance program.
   - develop quality control charts for each analysis done in the lab.

- review all analytical techniques, procedures, reagent preparation, and sample collection procedures annually.
- Be sure that everybody who performs laboratory testing participates in the QA/QC program.

3

Weights, Volumes
and Concentrations

A wastewater laboratory is precisely quantitative; solutions are prepared in concentrations of parts per million. Every milligram counts, and the results of measurements will depend upon the absolute accuracy of the chemical preparations.

Weights: Analytical Balances

The analytical balance is a delicate instrument with operation based on automatic internal setting of weights equivalent to the weight put on the pan—to achieve a "balance". It is extremely accurate, and enables weights prepared to the milligram (.001 g.).

The balance must be kept scrupulously clean. Corrosion from spilled chemicals will restrict the free movement of the pan and will work its way into the operating mechanism.

Seven Analytical Balance Troubleshooting Tips

1. Calibrate the analytical balance against standard gram weights often.
2. Do not weigh anything that is not room temperature. Cold substances will weigh more. Hot substances will weigh less.
3. Close balance doors when weighing. Drafts, uneven counters, or nearby bumps and bangs will disturb the weighing operation.
4. Begin by zeroing the balance.

5. Analytical balances are constructed for small weighing operations; do not overload it with a weight too near its maximum capacity.
6. Leave the balance clean, with the doors closed, when finished.
7. If the weight required is too small to weigh accurately on the analytical balance (under 10mg), prepare the solution in a greater concentration than needed, and then dilute it.

Desiccator

A desiccator is a glass container used to keep substances dry while they are cooling. It has a suspended floor, under which is a chemical desiccant which absorbs moisture from the air. Desiccant should be replaced periodically. Indicating desiccants change color when they are spent.
Desiccant chemicals: $CaCl_2$, $CaSO_4$, $NaCl$, silica gel.

Volumetric Measurements

Volumetric measurements are just as important as gravimetric ones. If 2.5416 grams of chemical is weighed out accurately and dissolved into a liter of water - which isn't really a liter because the volume wasn't measured correctly, the concentration of the solution is not going to be accurate.

Each piece of volumetric glassware is for a specific purpose. Beakers are meant only for mixing chemicals; they should not be used for measuring volumes. Beakers have the least calibration accuracy of all laboratory glassware. Erlenmeyer flasks are to be used for titration; the narrowed tops are designed to keep the liquid from splashing out of a swirling flask. These flasks are not accurate enough to measure liquids in. Volumetric flasks, graduated cylinders and pipets are the most accurate, and are meant for measuring liquid volumes in. Use these whenever exact measurements are indicated.

Pipets have several designations:
- "To Contain" means the pipet holds exactly the volume specified. This includes volumetric pipets and graduated pipets which are calibrated down to the tip.
- "To Deliver/blow Out" pipets are meant to be drained and then to be blown out so that all the liquid is ejected.
- "To Deliver" pipets should be drained, and the last drop should be drawn out by touching the tip to the side of the glass. Blow out technique should not be used with these.

Liquid chemical should always be room temperature when measuring. Water expands a great deal when it is hot, and later will contract, making the volume measurement deficient.

Choose glassware size based on common sense. To stir 500 ml of liquid, choose a 1 Liter beaker. When titrating into a 100 ml sample, put the sample into a 250 ml Erlenmeyer flask. To measure liquids 10 ml or less, use a pipet. Volumetric pipets are the most accurate. When measuring with a graduated cylinder use the size that is just a little larger than the volume of liquid needed. A buret is very accurate and can always be used to measure small amounts of liquids, but it is awkward. Burets are for titrating.

Read the Meniscus: Water molecules are attracted to the molecules of the container they are in, and the water surface will "climb up" the sides a little, forming a curve, a meniscus. Read the water level at the bottom of the meniscus. This applies to reading pipets, burets, graduated cylinders, volumetric flasks.

To prepare an exact concentration of a solution: Weigh the dry chemical on the Analytical Balance. Dissolve the chemical in a beaker with about half the amount of distilled water required. Pour into a volumetric flask which is calibrated to the right volume. Rinse the beaker out into the flask with a little distilled water (transfer quantitatively). Fill the flask to the calibration line with distilled water. Cap it, and Mix Well! Turn upside down at least 20 times slowly. It is difficult to mix liquids well in a full volumetric flask. Do not pour dry chemical down the neck of a volumetric flask!

Concentration

Concentration indicates how much solute is contained in a given amount of solution.

Parts Per Million

A weight to weight ratio, this means parts of the chemical or physical constituent, per million parts of water, by weight.

Take phosphorus, for instance: 10 parts per million
                               is 10 pounds P per million pounds of water
                               or 10 tons P per million tons of water
                               or 10 grams P per million grams of water
                               or 10 mg P per million mg of water

These are all the same concentration, and they are all weight/weight. Since by weight, a liter of water is composed of 1 million milligrams of water, then:

$$10 \text{ mg P per million mg water} = 10 \text{ mg P per Liter of water}$$
$$10 \text{ ppm} = 10 \text{ mg/L}$$

Micrograms Per Milliliter

In Standard Methods there is frequently stated a less familiar unit label, $\mu g/ml$. This is also a ppm designation, and is exactly the same as mg/L. It is sometimes used for designating small concentrations. To demonstrate how $\mu g/ml$ and mg/L relate to each other:

A microgram is a thousand times smaller than a milligram; if the unit label is written out as a fraction, and then numerator and denominator are multiplied by 1000 (not changing the value of the fraction), it becomes mg/L.

$$\frac{\mu g}{ml} \times \frac{1000}{1000} = \frac{mg}{L} \qquad \text{Therefore, } \mu g/ml = mg/L$$

Milligrams Per Kilogram

The unit label mg/kg is used widely within the environmental industry. It specifies the concentration of a solid within a solid, and is also a ppm designation.

Example: The concentration of copper in a chunk of gold, would be mg/kg. The concentration of mercury in a fish, would be mg/kg.

Molarity

This is the chemist's favorite way to state concentration. Molarity is used as the concentration label in a wastewater laboratory when preparing acids and bases. It employs the gram molecular weight of the compound.

1 Molar solution—the molecular weight of the chemical, in grams, dissolved in enough water to make 1 Liter.

With sodium hydroxide (NaOH) as an example:
First, calculate the molecular weight.

Na   23
O   16   Therefore, 40 grams NaOH, dissolved in 1 liter of
H   + 1   water = 1 Molar NaOH.
     40

Stated another way:
A 1 M solution of NaOH = 40 grams/Liter - or 40,000 mg/L

Normality

Used widely in the wastewater treatment industry, Normality is similar to Molarity; instead of gram molecular weight, gram equivalent weight is used. The gram equivalent weight is that weight of a compound that contains one gram atom of available hydrogen, or its equivalent. It is the number of grams of the compound that are required to react with one gram of hydrogen.

To work this out simply, pick a compound:

Example: NaOH   or $Na^+ (OH)^-$

Look at the positive ion. Count the charges. If it has one, then one molecule of it will react with one hydrogen ion (which also has one positive charge); it is one gram equivalent.

Therefore, for this compound, a 1 Molar solution = a 1 Normal solution.

40 grams NaOH, dissolved in 1 Liter = 1M = 1N

However, if the compound has two or more hydrogen ion equivalents, then the Normal solution concentration is calculated based on each of those equivalents.

Example: H_2SO_4 or $H_2^{++} (SO_4)^{--}$

In sulfuric acid, there are two positive charges (two hydrogen ion equivalents).

Calculate molecular weight:

2H   2
 S   32
4O +64
      98

98 g. H_2SO_4 dissolved in 1 Liter = 1M (one gram molecular weight)
98 g. H_2SO_4 dissolved in 1 Liter = 2N (two gram equivalent weights)
49 g. H_2SO_4 dissolved in 1 Liter = 1N (one gram equivalent weight)
or   1N solution H_2SO_4 = 49 g/L or 49,000 mg/L

Molarity is a matter of weights. Normality is a matter of weights and charges.

Liquids

In practice, however, these calculations are based on the assumption that the compound being used is a dry chemical, which can be weighed. HCl, H_2SO_4 are purchased in concentrated liquid form from the chemical distributor, and it is customary to measure a volume, not a weight. In addition, they are quite concentrated, and their liquid weight is greater per unit volume, than water. (density is greater—Specific Gravity is greater).

Example: For H_2SO_4, one ml. is not the same weight as one gram. One cannot measure out 49 ml, and expect it to be 49 g. Concentrated sulfuric acid is 1.83 times as heavy as water (SG=1.83).

To correct for this, divide the needed gram weight by the Specific Gravity.

49 grams needed to prepare 1N H_2SO_4. $\frac{49}{1.83}$ = 27 ml.

Measure out 27 ml and fill to 1 Liter with water.

Remember to add acid to water! (add acid to about half the volume of water in a beaker, mix, fill to 1 Liter total). To obtain the correct Specific Gravity for a laboratory chemical, look on the bottle label, or check the Handbook of Chemistry.
This type of calculation is needed in treatment plant use with liquid process chemicals, which are often shipped in concentrated form, and are denser than water (liquid alum, liquid ferric chloride). The Specific Gravity should be calculated in when determining chemical dosage rates.
There is one more factor to consider. In the laboratory, and for the treatment process, a purchased liquid chemical as shipped, is already diluted in water. It is not 100% pure. For instance, concentrated sulfuric acid is 96% acid in the chemical bottle. The other 4% is water. As calculated above, if this were 100% acid, 27 ml of it would be needed to prepare a 1 Normal solution. Since it is 96% pure, the required volume is

$\frac{27}{.96}$ = 28 ml.

Concentrated HCl is only about 37% acid; in this case the difference is very significant.

Concentration of Common Reagents

	HCl	HNO$_3$	H$_2$SO$_4$	H$_3$PO$_4$	CH$_3$COOH	NH$_4$OH
Specific Gravity	1.18	1.41	1.84	1.69	1.06	.9
Strength of Conc. Reagent	37%	70%	96%	85%	100%	29%
Molecular Weight	36.5	63	98.1	98	60	35.1
Molarity of Conc. Reagent	12	16	18	15	17	15
Number ml. Reagent to Prepare 1L of 1M Solution	83	64	56	70	42	66

For preparation of common acids and bases, see inside the front cover of Standard Methods.

Percent Concentration

Percent by weight concentrations are used when dealing with concentrated chemical solutions and process sludges.

$$\frac{1 \text{ gram chemical}}{100 \text{ grams water}} = 1\%$$

or $\quad \dfrac{1 \text{ gram chemical}}{100 \text{ ml. water}} = 1\%$

or $\quad \dfrac{10 \text{ grams chemical}}{1000 \text{ ml. water}} = 1\%$

or $\quad 10,000 \text{ mg/L} = 1\%$

To convert a 1N solution to percent concentration, first change it to mg/L, then to percent.

Example: $\quad$ 1N H$_2$SO$_4$ $\quad$ = $\quad$ 49 g/L

$\qquad\qquad$ or $\qquad\qquad\quad$ 49,000 mg/L

$\qquad\qquad$ or $\qquad\qquad\quad$ 4.9%

Changing Liquid Concentrations — (C X V = C X V)

The above formula is handy for dealing with laboratory chemicals or process chemicals; it can also be useful for combining process flows, and dewatering sludges; whenever liquid concentrations are changed. Any unit label for concentration, or for volume can be used.

The formula originated from Boyle's and Charles' Ideal Gas Law, $PV=nRT$. The pressure, the temperature, and the gas constant have been eliminated since they are the same for the solution before and after the concentration is changed.

There are several different ways to use this formula:

Example: The chemist has a .3N solution of HCl. How many ml of it would be needed to prepare 100 ml of a .01N solution ?

You Have			You Want to Get		
C	x	V	=	C	x	V
.3N	x	V(ml)	=	.01N	x	100ml
		V	=	3.3 ml		

Example: The chemist has 75 ml of a 1000 mg/L solution of NaCl. He wants to dilute this entire solution to a 50 mg/L solution. How much water will be needed for the dilution?

First calculate the amount of total solution that will result from making the dilution. This will be needed anyway to choose glassware size.

You Have			You Want to Get		
C	x	V	=	C	x	V
1000 mg/L	x	75ml	=	50mg/L	x	V
		1500 ml	=	V		

Subtract original solution volume to determine water needed:

1500 ml total volume - 75 ml original NaCl volume = 1425 ml water needed.

For any chemical procedure, use common sense. If the volume indicated to make the dilution is too small to measure accurately with a pipet, make serial dilutions. Dilute it first to an intermediate strength, then down to the concentration needed.

Ion Concentration

At times is desired to prepare a solution which has a specific concentration of a single element. The chemist must be able to calculate the amount of compound needed, so that he will end up with the correct concentration of the desired element.

Example: It is desired to make up a solution which contains 100 mg/L of sodium. The compound available is NaCl.

First, calculate the % weight that sodium is, in this compound:

$$\begin{array}{ll} \text{Na} & 23 \\ \text{Cl} & +35 \\ \hline & 58 \end{array} \qquad \frac{23}{58} = .4 = 40\% \quad \text{The compound is 40\% sodium.}$$

Then divide the desired concentration by the % sodium in this compound.

$$\frac{100 \text{ mg (sodium desired in the liter)}}{.4} = 250 \text{ mg}$$

Weigh out 250 mg NaCl; dissolve in 1 Liter - it will be 100 mg/L Na.
When labeling this solution, it would be called 100 mg/L NaCl, as Na.

This same procedure can be used to calculate the concentration of any part of a molecule in a liquid solution. It is particularly useful with nitrogen ions, and has become accepted to refer to these in terms of the weight of nitrogen in the compound.

Example: A solution which contains 100 mg/L $NaNO_3$, as N means that the compound in use is sodium nitrate, but 100 mg/L is the concentration of nitrogen in this solution.

Laboratory Apparatus:
Chemicals and Instruments

Chemicals

The Chemical Label

Know your chemicals! When preparing reagents, always read the label on the chemical bottle. Anyone can make a mistake and pull the wrong chemical off the shelf.

The chemical name is on the front. Read it carefully. Some sound very similar (ex. sodium sulfate and sodium sulfite are very different chemicals). The chemical state is often under the name (powder, granular, liquid). If it is a liquid, the concentration is given. The exact chemical formula and the formula weight are on the left side. Be sure to compare against the formula required in the test procedure. Under the formula is the assay. It gives a listing of impurities mixed with this chemical, and their percentages.

On the bottom of the chemical label are warnings:
- Danger!
- Poison!
- Special Handling notes

Some chemical distributors have adopted the National Fire Prevention Association safety hazard color codes for labeling stock chemicals:
- *Gray*—moderate hazard
- *Blue*—health hazard; toxic if inhaled, ingested, absorbed
- *Yellow*—reactive; oxidizing agent
- *White*—corrosive
- *Red*—flammable

For most analyses, ACS (American Chemical Society) Reagent Grade chemica
are satisfactory except in certain instances where ultra high purity chemicals a:
required. Any chemical which has been removed from the chemical bottle shoul
be used, or disposed of. Don't put it back into the bottle. To find out more abo·
a chemical, check with the <u>Merck Index</u>, or the <u>Handbook of Chemistry</u>. Standaı
<u>Methods</u> also has useful information.

Chemicals and Water

Some dry chemicals are labeled "Anhydrous". This means that the chemic.
formula is that of the compound only (ex. NaCl may be labeled Sodium Chloridı
or Anhydrous Sodium Chloride, but either way it is just NaCl).

Some dry chemicals are hydroscopic and will readily absorb moisture from tł
air. This is just physical moisture, and doesn't change the nature of the chemica
or the formula. However, when preparing a solution using a specific amount ı
this chemical, some of what is being weighed out is the moisture, and there is r
way to know how much of it that is. The chemical solution will end up beir
weaker than intended. This type of chemical must first be dried in an oven to driv
off the water. Most lab procedures anticipate this problem, and include directioı
to dry certain chemicals for a stated amount of time before weighing. Do nı
exceed the required temperature, or the chemical may begin to decompose.

Some chemicals have chemically bound water attached to each molecule (e:
Manganous Sulfate, $MnSO_4 \cdot 4H_2O$). This may be labeled by the compound nanr
only, or it may be labeled Manganous Sulfate-4 Hydrate. In any case, as verifiı
by the formula, this compound has four water molecules attached to eveı
manganous sulfate molecule. This is not just a moist chemical. It is dry, but watı
is a permanent part of the molecule. Drying this chemical in an oven at norm.
chemical drying temperatures will not take this water off. Heating at much highı
temperatures might, but then the compound would be decomposing.   This
chemically bound water; it has weight, and that extra weight, if not accounted fo
can throw the solution concentration off by a great deal. However, this is
reliable, constant weight, and it can be calculated.

For Example, if the test procedure required preparing a solution with 10 gran
of MnSO4, more than 10 grams would be needed if the chemical chosen weı
$MnSO_4 \cdot 4H_2O$.

First calculate the percentage of MnSO₄ in this chemical.
Add up the atomic weights of each element.

$MnSO_4 \bullet 4H_2O$		$MnSO_4$	
Mn	55	Mn	55
S	32	S	32
4O	64	4O	+64
8H	8		151
4O	+64		
	223		

$$\frac{151}{223} = .68 = 68\% \text{ of this chemical is MnSO4.}$$

Now divide the weight needed by that percentage:

$$\underline{10 \text{ grams needed}} = 14.7 \text{ grams of the chemical must be weighed out}$$
in order to get 10 grams of $MnSO_4$.

Distilled Water

The quality of reagents is strongly dependent upon the quality of water used to prepare them. There must be no detectible concentration of the substance to be analyzed in the water. For use in a laboratory where multiple analyses are being performed, a source of the purest water available is needed: distilled water. Check it monthly by measuring the electrical conductivity (pure distilled water does not conduct electricity). EPA requires that distilled water in a certified water testing laboratory measure <1 umho/cm at 25 degrees Celsius.

However, contaminants do enter. The system is open to atmosphere; carbon dioxide and oxygen enter. If the collection bottle is not completely clean, nutrients will dissolve, and algae will grow in the bottle (especially if it's near a window). As algae die, they provide food for bacterial growth. Dirty valves and delivery tubes are prime causes of sample and reagent contamination. Collection bottles should be acid washed periodically. They must be kept clean, and closed. Distilled water does not maintain original purity over an extended period of time. It is best not to plan to store great quantities of it; it should be prepared as needed, and used.

Chemical Indicators

Indicators are organic chemical dyes which react to produce a color under certain conditions. They are visual aids, used to indicate the end point of titration.

Two types of indicators are most widely in use in wastewater labs:

Adsorption Indicators: When added to the sample, these chemicals loosely bind up a few molecules of the compound being measured, and a color change is noted. As the titration nears completion, the titrant demands these molecules back from the indicator, and the color is discharged, marking the end point. For example, starch is an adsorption indicator. We use it when titrating for dissolved oxygen; a few drops of starch produces a strong blue color in combination with the sample molecules. This color disappears immediately as those last molecules are taken by the titrant. This color change indicates that the titration is completed.

pH Indicators: pH indicators change color at specific pH's. They are used for titrations which are performed to neutralize an acid or alkaline component in the sample.

Frequently used are:

Phenolphthalien: colorless (titrating up, with a base)
--------------------------> pink: at pH 8.3

Methyl Orange: orange (titrating down, with an acid)
--------------------------> rose: at pH 4.5

How much to use? Directions for use of indicators usually specify "a few drops". This is a flexible amount. Use enough of it to obtain a decent color to look at, and to notice a change in.

Standards

A Chemical Standard is a solution which contains a known concentration of the chemical component of interest.

Standards are used to calibrate instruments. For instance, a pH meter is made to read correctly two solutions of known pH, standards (for pH meters they are more commonly called "buffers"), before attempting to read the pH of the sample.

Standards are in use for titration. A titrant must have an exact concentration. It is the reagent which measures the component being tested for. A Primary Standard is a chemical solution which, as prepared, can be relied upon to be pure and to be the exact concentration it is made up at. A Secondary Standard is a solution which is prepared from an impure chemical, and its exact concentration was not known. It is reacted with a primary standard (standardized) on a mole to mole basis, to determine its exact concentration. Most often it is the secondary standard that is the needed compound as a titrant for the test procedure; the purpose for the primary standard is to standardize the secondary.

Standards are in use in colorimetric analysis. They are prepared as a series of solutions with increasing known concentrations of the constituent to be analyzed. A reagent is added to each which produces visible color, increasingly dense for the more concentrated standards. By adding the reagent to the unknown sample also, its color can be compared against that of the standards to determine concentration.

Stock, Standard, & Working Standards: In a colorimetric lab procedure which tests for very tiny amounts of a contaminant in a water sample (ex. 2 mg/L expected), standards must be made up in very dilute concentrations. To minimize weighing problems and to increase shelf life of the solution, it is often practical to first prepare a much stronger Stock Solution (ex. 1000 mg/L), and then dilute it down to a Standard Solution (ex. 10 mg/L). From this Standard Solution, the chemist chooses his Working Standards, those that he will actually use for the color comparison with the sample (ex. 1mg/L, 2mg/L, 3mg/L, 4mg/L). The Working Standards are prepared in a concentration range that will straddle the sample concentration. A prepared standard which has 0 concentration of the substance to be analyzed, is called a Blank.

Chemical Measurement Instruments

Since a wastewater lab is an instrumental lab, it is important to understand some of the working mechanism and the principles upon which the instruments were developed and operate. No instrument functions perfectly forever. Proper maintenance, replacement of necessary components and troubleshooting of some of the more obvious problems that occur is part of the chemist's responsibility.

The pH Meter

pH is defined as the negative log of the hydrogen ion concentration of the solution. This is a measure of the hydrogen ion concentration of the solution. In short, it is the relative acidity or basicity of the solution.

The chemical and physical properties, and the reactivity of almost every component in water are dependent upon pH. Its measurement is probably the most widely used electrical method of analysis in a chemistry laboratory. This is a potentiometric method - based on a voltage reading. The meter is a voltmeter, and reads millivolts of electricity.

Working Mechanism: The sensing mechanism consists of two electrodes, a Reference Electrode and a pH Electrode; both must be immersed in the sample for a proper reading.

Reference Electrode: The reference electrode provides a constant voltage to the meter, which acts as a standard against which the sample reading is compared. The voltage is provided by a flow of electrolyte filling solution, a pH 7, saturated KCl solution (4M), from the electrode into the sample through a small diffusion port near or at the bottom (notice a small, round, ceramic "clogged hole"). Clogging of this ceramic junction is the most frequent cause of electrode failure; if it fouls, follow the manual's cleaning procedure; otherwise replace the electrode. The reference electrode is filled through a small opening near the top. and it must be kept full. The filling port has a sleeve or cap on it, which must be kept closed when not in use, to prevent evaporation of the electrolyte, and open when in use, allowing air flow.

Types of Reference Electrodes: A Calomel reference electrode has Hg/HgCl internal wires. This kind is used for most routine applications; it is easy to use and maintain, and uses saturated KCl electrolyte; shelf life is 2-3 years. A Silver/Silver Chloride reference electrode has Ag/AgCl internal wires. This is used for high temperature applications, and uses saturated KCl + AgCl electrolyte, to prevent dissolving the internals. A Double Junction reference electrode is a special double body type meant for use with very dirty samples, and with industrial samples which may cause interference due to a component in the sample. A different filling solution can be used. The Gel-type reference electrode is made with a solid filling solution—a permanent gel—not to be refilled. It does diffuse out into the sample slowly through the diffusion port—just as the others do. The electrode is used till the gel is gone (approximately one year); then the electrode is replaced.

pH Electrode: This is a glass membrane electrode which is actually sensing the pH of the sample. Hydrogen ions diffuse through the membrane to the inside of the electrode, where a wire picks up the reading and records the voltage carried in from the sample. This will be a different voltage reading than that coming in from the reference electrode. The meter records the difference between them, in millivolts. It is then converted to pH.

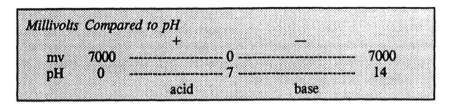

Combination Electrodes: A combination electrode is just what it says, both electrodes in one probe. The reference electrode is the outer portion with filling port and bottom diffusion port. The pH electrode is the inner portion, which extends all the way down to the very tip, which is a glass membrane. It is best to have an open plastic shield over the end of the probe to protect the membrane from breakage. The advantage of the combination electrode is that only one probe is immersed in the sample. Polymer body combination electrodes (plastic), are made for rough handling applications, and are available in liquid filled or gel filled versions.

Meter Calibration: pH meters are calibrated (set to a correct reading) with purchased solutions which have specific pH values. They are called Buffers, and are composed of phosphates, acetates, etc. depending upon the pH of the buffer. These are the standards with which the meter is set to work correctly. The pH is listed on each buffer bottle, and it is dependable. Buffers can also be purchased in dry capsule form (long shelf life) and made up into solutions as needed.

If using a newer pH meter, follow the instructions carefully on the meter for calibration. Instructions are somewhat different for each meter. Use the buffers which are indicated for calibration with that meter. Always calibrate a pH meter with two buffers. A pH meter, near the end of calibration, will usually give a reading of "slope". This has to do with the efficiency of the probe. It means millivolts sensed, per decade of concentration (to pH, a decade of concentration is the difference between pH units, since they multiply by ten from one to another). The instrument divides these mv/decade concentration by the ideal mv/decade concentration and obtains a percentage (of probe efficiency). The reading should be over 90% slope; some meters will adjust for the slight amount off 100% accuracy. If the slope reading is under 90%, do the calibration over; if still off, recondition or change the probe.

If using an older meter, and there are no directions for calibration, choose two buffers which straddle the expected pH of the sample. Immerse the probe in one of them, and set the calibration knob to read that pH. Then do the same with the other buffer. Then set the meter to "read pH", and read the sample.

The Buffer solution is really a "buffer". It neutralizes any possible acid or alkaline influences present in the water in which it is dissolved, and the pH should reside very stably at its designated value. When the pH probe is immersed in the buffer, the needle should not drift. If it does, recondition or change the probe. A pH meter operates in millivolts, and it will give millivolt readings all the way through calibration, making the conversion to pH units at the end.

Fifteen pH Troubleshooting Tips

1. Open air port when using. Close when done.
2. Rinse between samples with distilled water.
3. Do not be concerned with odd readings obtained while probe is rinsing in distilled water. Although distilled water should have a pH of 7 at 25 degrees Celsius, the probe has difficulty reading anything at all in distilled water, which does not conduct electricity.
4. Handle probe carefully—glass membranes break easily.
5. Wait for sample reading to settle down before recording it. Wait 5-6 minutes before taking a reading with any probe method (cleaner waters often have low ionic strength and a slow response time).
6. If an automatic temperature adjustment probe is included with the meter, this must be immersed in the solution with the pH probe, in order to sense the temperature and read the correct pH.
7. Needle drift is caused most frequently by the sample changing—check for drift when immersed in pH 7 buffer. If there is drift when in buffer, change the filling solution.
8. Don't stir the solution with the pH probe; it may cause the needle to swing. However, if the sample is not stirred at all, a polarity may build up around the probe. It is best to stir very slowly with a stir bar.
9. Don't immerse in very high or low pH solutions. This shocks the probe; it may have diffculty recovering.
10. Don't immerse probe in sample deeper than level of the electrolyte. Sample will flow in, rather than electrolyte flowing out.
11. Probe contamination from electrolyte crystallization can be rinsed off with distilled water. See conditioning instructions in manufacturer's pamphlet.
12. If readings are poor, it is more likely the probe than the meter that is at fault.
13. Troubleshoot the obvious first:
    *   Is it plugged in?
    *   Is there electrolyte in the probe?
    *   Is the meter switched to "pH"?
    *   Is the glass membrane broken?
    *   Is the sleeve on the filling port open?

14. After use, store probe immersed in pH 7 buffer. Never let a pH probe go dry.
15. Keep probe clean between uses. pH 7 buffer is phosphate; it stimulates algae & bacterial growth; change frequently.

Ion Specific Electrodes

Measurement by Ion Specific Electrode (ISE) is a fast growing technology, and is applicable to almost every laboratory, and most dissolved ions.

Its chief advantage is quick results; this is a rapid method of measuring for dissolved ions in the sample with a minimum of preparation time, reagents, space and equipment.

For most measurements, other substances or ions in the sample do not interfere with the reading. This is a definite advantage over colorimetric methods.

The chief disadvantage is cost; most electrodes have a usable life of one year, and should be replaced annually.

These are potentiometric measurements. Any pH meter which reads out in millivolts can be used with Ion Specific Electrodes (the measurement of pH is actually an ion specific electrode method; it is measuring hydrogen ions). In municipal and industrial wastewater testing laboratories, it is common to see the measurement of ammonia, nitrate, and chlorine residual performed by ion specific electrode. Since there are some differences from one electrode to another, more specific details are included in the test procedure chapters.

Colorimetry

Colorimetry began many years ago when chemists first realized that certain dissolved constituents in water, when added to specific reagents, will create a reproducible visible color, and that the density of color will be directly proportional to the concentration of that constituent in the water.

Beer's Law: August Beer, in 1850, proposed the theory that the amount of light absorbed or transmitted through a colored solution, will be directly proportional to the density of color of the solution. This is known as Beer's Law and has been the basis of colorimetric analyses ever since.

It was also noted, by Johan Lambert, that the absorbance or transmittance of light through a colored sample is directly proportional to the width of the sample. This is Lambert's Law. In modern colorimetric procedures, we hold this factor steady by using the same size sample containers for all samples.

With this knowledge, the chemist was able to produce color in a normally colorless sample, in order to identify the presence of a contaminant, and to distinguish between a more and less concentrated contaminant. In 1852, for ammonia determinations in water samples (ammonia was at that time the prime indicator of polluted water), the use of Nessler Tubes was established. A series of increasingly concentrated standards was made up in test tubes, each containing a known amount of ammonia. When the color reagent was added, and color developed, the colors ranged from lighter to darker along the line of test tubes. The unknown sample, plus color reagent, developed a density of color which fell somewhere in among the standard colors, and its ammonia concentration was determined visually. With this, colorimetric methods were begun, and the use of colorimetric standards was established.

Color Comparison Tubes were the first sets of colorimetric standards in widespread use; they are still considered adequate for rough comparison, and are used in the field where instruments are not available. Permanent sets are available in a range of colors. When necessary, some chemists will make them up specifically to suit a project. Using these permanent colors eliminates the task of making up a whole set of standards each time a test is run, but there is a risk of losing accuracy if the color hue of the standards is not comparable to that of the developed color in the sample, or if the color producing reagent is not prepared in exactly the same way each time.

The Colorimeter: Improvements were made. To eliminate the tubes, the colorimeter was developed. Small, manually operated models are still found in some wastewater treatment plants. The sample, with color producing reagent added, is placed into a glass vial, a cuvette, inserted into a small box, and held up to the light for viewing. A series of translucent glass or plastic colored discs arranged on cards, or on a color wheel, and labeled with their respective equivalent concentrations, could be held up alongside the sample to match the colors. Sets of different colors, reds, yellows, blues, for identification of separate constituents, could be purchased. Colorimeters gained popularity as a quick test method for many water constituents. Advantages here over color comparison tubes were in eliminating interference from varying daylight intensity and reducing equipment handling.

The major disadvantages of these early colorimetric methods were that the permanent color solutions or the discs did not always match the sample colors well, and that the use of daylight, which is of different intensity each day, and throws other colors into the sample, resulted in a loss of accuracy. In addition, as each new batch of color producing reagent was prepared, if there was a slight change in its capacity to develop color, the accuracy of the sample color would be decreased. Finally, human eyesight, a very imperfect measuring instrument, was still being used to read the colors.

The Spectrophotometer: By 1940, the automatic colorimeter, or spectrophotometer, had been invented. An incandescent light is the source, which shines through a colored filter and deletes all wavelengths of light (all colors) except those which will be most strongly absorbed by the sample's color. That portion of light which passes through the filter then passes on through the sample. Some of it is absorbed, but that which is transmitted through the sample is sensed by a photocell, which converts it to a voltage. This is measured by a voltmeter, and then recorded on an analog scale (needle reading) as % Absorbance, or % Transmittance. More recent improvements include a prism inserted just beyond the light source to break up the light into a rainbow of colors, separating all the wavelengths so that each color can be picked up individually. The filter is replaced with a movable opaque plate which has a vertical slit in it. The chemist moves this plate back and forth, choosing the wavelength, till it resides in front of the exact wavelength of light he wishes to allow through to the sample. This provides maximum choice among visible wavelengths of light.

There may be several wavelengths of light that a given colored sample will absorb. The one at which it is absorbed most strongly will give a greater reading, or "peak". For a given color, there may be two or more good peaks from which to work. The range of visible light on the electromagnetic wave spectrum is a small band from 400-700 nanometers (nm—the unit label for wavelength). It is within this range that cells in our eyes can pick up electromagnetic waves, and register it to the brain as color. It is now possible to enter the ultraviolet or infrared zones with the spectrophotometer by using a different light source, one which will emit that type of light. However, range is limited with these. The instrument has been designed for the visible light range; instrument materials, and the surrounding air may interfere.

According to Beer's Law, absorbance or transmittance of light can be sensed and recorded by the instrument. The spectrophotometer is first zeroed to "zero absorbance", or to 100% transmittance. The absorbance mode is most frequently used in wastewater chemistry. Transmittance is used when the visible color is so slight that it will not absorb much light, or when the graphing of absorbance will be inverted.

Operation: For maximum accuracy, colorimetry has come full circle back to the use of a set of standards, prepared from the chemical constituent to be measured. to these, along with the sample, is added a color producing reagent. Color is developed in all of them, and they are all read on the spectrophotometer. In this way, all the colors are of the same hue, there are no interfering colors given off from the light source, and a very precise and sensitive instrument is doing the reading.

Reagent Blank: The first step in operating a spectrophotometer is to zero the instrument. It is desired that the color sensed by the instrument be caused only by the developed color produced by the chemical being measured. Any color present in the reagent itself should be subtracted. To do this, prepare a blank made of distilled water plus the reagent, and zero the instrument to this. Do this every time a test is run. It is really a standard with 0 concentration of the measured constituent.

Calibration Curve: The instrument will yield a numerical reading for each of the standards, and for the sample, but that is a reading of % absorbance, or % transmittance, on a scale of 0-1, or 1-10. It is not a concentration value, for either the standards or the sample. The concentrations of the standards are known, but to calculate the concentration of the sample, a calibration curve is created from the standard readings, a slope is calculated, and from that slope the sample concentration is obtained.

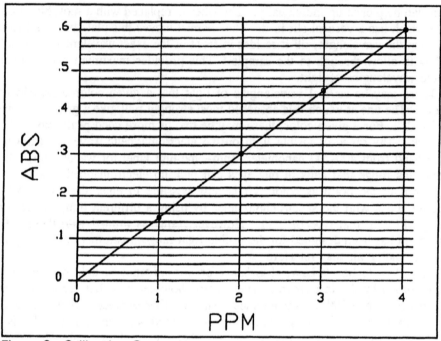

Figure 2 - Calibration Curve

Water utility personnel set up the graph like the one shown above, with standard concentrations in mg/L across the bottom, and Absorbance up the side. A dot or x is entered at each absorbance value for each standard, after it is read on the instrument. A standard slope is then created by throwing a straight line from the zero point on the graph, midway between the dots which designate the standard readings. Often the dots do not fall exactly into a straight line. Do not draw dot-to-dot. Throw a straight line through them. This curve must be straight; it is an average. EPA requires that at least three standards must be run to prepare a true calibration curve. If the sample absorbance reading falls outside the range of standards, officially the test is invalid.

Calculating Sample Concentration: There are two methods to obtain sample concentration:

1. Create the calibration curve from the standard readings. Enter the sample reading manually from the left, at its correct absorbance; slide over to the curve, then straight down to concentration. This is the concentration of the sample. This is exactly how sample and standard

color were compared manually to Nessler Tubes, by holding the sample tube up against the standards to see which one matches; using the instrument to make this comparison is much more accurate. Running the standards to prepare the calibration curve needs only to be done the first time the test is performed, and each time a batch of fresh reagent is prepared, but the graph will have to be at hand to read sample concentration each time a sample is tested.

2. After graphing the standards, calculate a numerical slope which will be representative of the slant of the standard curve:

$$slope = \frac{x \; mg/L}{y \; Abs.}$$

Note that the formula is inverted from the familiar arithmetic slope (y/x). This doesn't matter. It still designates the slant of that line; this method came about because the water industry has chosen to place mg/L on the bottom of the graph, and Absorbance up the side. Once the slope value has been calculated, it will apply equally to the standards and to the sample. Plug it back into the formula to calculate the concentration of the sample. For convenience, rearrange the formula to read:

Sample Absorbance x slope = Sample Concentration

The first time the test procedure is run, prepare and run standards, sample and reagent blank. Create calibration curve and calculate slope. After that, there is no need to refer to the graph each time the analysis is performed. Simply run sample and one standard (to verify that it's reading hasn't changed). Use the slope number to calculate concentration. The entire graphing procedure should also be done each time a new batch of reagent is prepared. A slight change in the nature of the reagent will change the graph, and the slope number.

Interferences: Turbidity in the sample interferes with colorimetric measurements. The spectrophotometer cannot distinguish turbidity from color, and a turbid sample will yield a false high reading. Filter the sample first with a .45 micron filter. In addition, a sample which is naturally colored will produce a false high reading. The color read by the instrument should originate only from the constituent/reagent complex, not from an extraneous natural color in the sample. Take a separate reading of the filtered sample with nothing at all added to it, and subtract this reading from the sample reading. If the standards have any color before the reagent is added, that must also be subtracted. To avoid this problem, choose a standard which has no color when dissolved.

Each colorimetric test procedure for chemical analysis has been developed for a specific range of concentrations. Its accuracy is dependent upon obtaining a straight line slope at that concentration. Attempting to run a test using standard concentrations well out of the range for which the test was developed may end up producing a truly curved slope (which the graphing procedure attempts to force into a straight line).   If the sample to be tested is out of range of the standard concentrations normally used for the test, it is better to dilute the sample than to raise the standard concentrations.

Solids

Solids in water are defined as any matter that remains as residue upon evaporation and drying at 103 degrees Celsius. They are separated into two classes: *suspended* and *dissolved*.

Total Solids = Suspended Solids + Dissolved Solids
(nonfilterable residue) (filterable residue)

Each of these has Volatile (organic) and Fixed (inorganic) components which can be separated by burning in a muffle furnace at 550 degrees Celsius. The organic components are converted to carbon dioxide and water, and the ash is left. Weight of the volatile solids can be calculated by subtracting the ash weight from the total dry weight of the solids.

DOMESTIC WASTEWATER

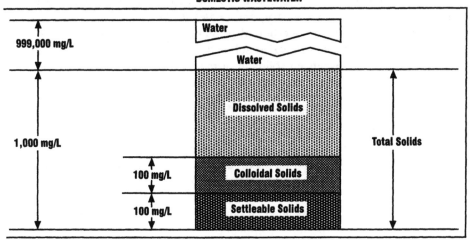

Total Suspended Solids/Volatile Solids

The Total Suspended Solids test is extremely valuable in the analysis of polluted waters. It is one of the two parameters which has federal discharge limits at 30 ppm through enforcement of the Clean Water Act. Solids are removed throughout the treatment plant to prevent excessive solids discharge to the receiving stream, which would contribute to lowering of dissolved oxygen available to life in the water, and to eutrophication of the stream.

Treatment Plant Significance

Solids determinations are very important in evaluating the performance of wastewater treatment plants, and in controlling the processes in the plant. Total Suspended Solids is performed most often on Raw Wastewater, Primary Effluent, and Final Effluent—at secondary treatment processes. A composite sample is taken for the test.

Raw Wastewater: Entering the treatment plant, due to water velocity, all particulate solids are suspended. Their concentration in the water determines design of process units, degree of treatment needed, and indicates changes in influent water quality. The volatile content of these suspended solids relates to the efficiency of pretreatment units, and indicates changes in the amount of organic content of the incoming wastewater. It will influence percent removals and control of secondary treatment, frequency of primary sludge pumping, and the efficiency of the anaerobic digester.

Primary Effluent: Total Suspended Solids concentration of this water determines the load on secondary, and relates to the efficiency of primary treatment. Frequency and duration of primary sludge pumping will have a significant effect on the suspended solids content of the primary effluent. Volatile content of suspended solids in primary effluent is composed of the particulate organic compounds that will be used directly by the organisms in secondary for growth, to be removed through biological adsorption, and then absorption into the bacterial cell. It will determine the bacterial growth rate, and effect aeration and mixed liquor concentration.
Return and waste activated sludge pumping rates will be affected. Non-volatile particulates will also be adsorbed by the biomass, and separated from the water, but these do not aid in bacterial growth.

Final Effluent: This suspended solids test is most important, for it determines NPDES permit compliance. Total suspended solids concentration of final effluent will depend on the efficiency of secondary treatment (adequate aeration, proper F/M & MCRT, good return and wasting rates), and the settling capacity of the secondary sludge. Excessive suspended solids in final effluent will adversely affect disinfection capacity. Volatile component of final effluent suspended solids is also monitored. A low volatile percentage may indicate hydraulic overload. High volatile percentage may indicate an unusual industrial input and may indicate a higher BOD reading; comparison must be made with other wastewater test results to determine the problem.

Expected Total Suspended Solids:
   Raw Domestic Wastewaters: 200-400 ppm (60-80% Volatile)
   Wastewater Secondary Effluents: <30 ppm (60-80% Volatile)

Treatment Plant Control

Problem: Higher than normal Total Suspended Solids concentration in final effluent

Possible Cause	Test This Sample Also For	Other Checks
Industrial high strength input	COD, Tot.Phos., NH₃-N, DO	IPP records Previous Operating Reports Contact contributing industries Raw & Prim.effluent COD, Tot.Phos., NH₃-N, TSS/VS, DO Primary sludge pumping rate, frequency DO, blower use in biological unit Sec.sludge blanket level
Industrial toxic shock	COD, NH₃-N, DO	Plant operating records Contact contributing industries Raw pH, color, odor DO, blower use in biological unit Biomass microscopic examination Floc formation Settleability of secondary sludge Sludge TS conc., blanket level
Inplant recycle input	COD, Tot.Phos., NH₃-N, DO	Recycle flow records Recycle COD, Tot.Phos., NH₃-N, TSS/VS, TS Plant flow records DO biological unit Sec.sludge blanket level

Possible Cause	Test This Sample Also For	Other Checks
Hydraulic overload	COD, Tot.Phos., NH_3-N, DO	Plant flow records Influent water temp. Raw & Prim.Eff. COD, Tot.Phos., NH_3-N, TSS/VS, DO Prim.sludge TS conc. Sec.sludge blanket level, TS conc.
Inplant operational control problem	COD, Tot.Phos., NH_3-N, DO	Coagulant feed rate, conc., type Jar Test Chemical feed pump operation Prim.sludge TS conc. Prim.Eff. TSS conc. Sludge pumping frequency Mechanical: sludge collector Mixed liquor conc. DO in biological unit Mixing in biological unit Sludge TS conc., blanket level Biomass microscopic examination F/M ratio, MCRT, RAS rate, hydraulic loading
Incorrect analysis		QC procedures Refer to Standard Methods

Analysis: Total Suspended Solids/Volatile Solids

Quality Control:
- Mix the sample well; pour aliquot before it gets a chance to settle. Use graduated cylinder to measure sample volume. Transfer quantitatively to Gooch Crucible (wash out any particles that are stuck on the inner walls of the cylinder). If large, uncharacteristic pieces are in the sample, remove them.
- Be sure the analytical balance has been calibrated to standard weights - recently. Weights to the milligram are necessary.
- Do not handle solids sample ceramics with fingers. Use tongs.
- Weigh only room temperature samples; allow warm samples to cool in a desiccator.
- To troubleshoot weight discrepancies, run a blank, using distilled water, along with the samples.
- As per Standard Methods, periodically dry, cool, weigh repeatedly till weight is constant. In this way, proper drying time can be established.
- Oven and furnace drying and burning temperatures are critical. A calibrated thermometer should be permanently set into the drying oven. Periodically it should be checked with another calibrated thermometer.

Don't allow the muffle furnace go over 550 degrees Celsius. Components in the filter paper may ignite; some inorganics in the sample may burn. At higher temperatures, some inorganics may burn.
- Do not open the furnace door after a sample has been placed in there for ignition. The extra oxygen that enters when the door is opened may ignite it so forcefully that the sample gets blown out of the crucible.

Apparatus:
- drying oven
- muffle furnace
- Gooch crucibles
- desiccator
- glass fiber filters
- source of vacuum
- filtering flask
- tongs

Procedure:
1. Insert filter with rough side up in Gooch crucible.
   Rinse filter with distilled water, applying vacuum to seat.
2. Dry filter & Gooch at 103 deg.C for 1 hr.
   Burn in muffle furnace for 15 minutes.
   Cool in desiccator.
3. Weigh Gooch/filter to the nearest mg.(tare weight)
4. Mix sample well so that aliquot used has a representative amount of solids. Transfer quantitatively; filter a known volume of sample (it is best to use the largest volume possible that will not blind the filter).
5. Dry in 103 deg.C oven for 1 hour. Cool in desiccator. Weigh again (dry weight).
6. Burn in muffle furnace for 15 minutes (550 deg.C). Cool in desiccator. Weigh again (ashed weight).
7. *Calculate:*

$$\text{TSS (mg/L)} = \frac{\text{dry weight(mg) - tare weight(mg)}}{\text{ml. sample}} \times 1000$$

$$\text{Fixed Solids (mg/L)} = \frac{\text{ashed weight(mg) - tare weight(mg)}}{\text{ml. sample}} \times 1000$$

$$\text{Volatile Solids (mg/L)} = \text{TSS (mg/L) - Fixed Solids (mg/L)}$$

$$\text{\% Volatile Solids} = \frac{\text{Volatile Solids (mg/L)}}{\text{TSS (mg/L)}} \times 100$$

Mixed Liquor Total Suspended Solids

This is similar to the Total Suspended Solids test done on influent and effluen wastewaters; it is done routinely on mixed liquor and activated sludge at an Activated Sludge Treatment Plant. Normally, Total Solids tests are done on sludges, but this is a less concentrated sludge, and the weight of the dissolved substances in it may be significant, so it is tested for solids in the same way as a wastewater sample is.

Treatment Plant Significance

The results of this analysis represents the weight of the organisms in the aeration tank, the bacteria which remove the organic wastes from the water. These bacteria are settled out in the secondary clarifier and called Activated Sludge. Continually some portion of this sludge is recycled back to aeration (Return Activated Sludge) and the rest is wasted (Waste Activated Sludge).

The Mixed Liquor is a suspended floc mass, and includes not only bacteria, but also the adsorbed solids from the wastewater, as they are being digested by the bacteria. The Mixed Liquor Suspended Solids (MLSS) concentration is controllable by varying the RAS and WAS pumping rates, and this concentration will determine aeration rate, F/M ratio, sludge age & MCRT. Human control over the biomass concentration is a major advantage of operating an activated sludge treatment plant. Volatile and Fixed percentages of the mixed liquor are used to determine biological efficiency and potential for waste removal.

> Expected MLTSS concentration: 1500 - 4000 mg/L
> Expected TSS Activated Sludge: 5,000 - 30,000 mg/L (.5-3%)

Treatment Plant Control

Problem: Mixed Liquor Total Suspended Solids other than normal concentration.

Possible Cause	Test This Sample Also For	Other Checks
Industrial high strength input	DO, TS	Prim: Raw & Prim.Eff. COD, NH$_3$-N, Tot.Phos, TSS/VS Blower use RAS, WAS rate Sludge TS conc.

Possible Cause	Test This Sample Also For	Other Checks
Industrial toxic shock	DO, pH microscopic exam	Raw pH Floc formation Settleability of Sec. sludge
Inplant recycle input	DO, TS	Recycle flow records Recycle COD, Tot.Phos., NH$_3$-N, TSS/VS, TS. Plant flow records Sludge blanket level
Hydraulic overload	DO, TS	Plant flow records Influent water temp. Raw & Prim.Eff. COD, Tot.Phos., NH$_3$-N, TSS/VS, DO Prim.sludge TS conc. Sludge blanket level Sludge TS conc.
Inplant operational	DO microscopic exam	Coagulant feed rate, Mixed Liquor conc., type Jar Test Chemical feed pump operation Prim.sludge TS conc. Prim.Eff. TSS conc. Sludge pumping frequency Mechanical: sludge pump, sludge collector Mixed liquor conc. Sec.sludge TS conc. Sludge blanket level F/M ratio, MCRT, RAS rate, hydraulic loading
Incorrect analysis		QC procedures Refer to Standard Methods

Analysis: Mixed Liquor Total Suspended Solids

Grab samples are taken for this test. This is a Total Suspended Solids test, but because of the greater amount of solids, a Buchner Funnel is used for filtering the sample. If Volatile Solids is to be determined, the filter paper must be glass-fiber. The filter paper (plus solids) is removed from the funnel, dried in the oven, weighed and then burned, and weighed again.

Quality Control:
- Mix sample thoroughly, and pour immediately!
- Be sure to dry the sample thoroughly. To check time needed, dry, cool, weigh, dry again, cool, weigh.

- See Quality Control and Safety notes from Total Suspended Solids section.

Apparatus:
- drying oven
- desiccator
- tongs
- filtering flask
- source of vacuum
- buchner funnel
- filter paper with diameter at least 1 inch larger than funnel base

Procedure:
1. Dry filter paper in oven at 103 deg.C for one hour. If volatile solids weight is desired, burn in muffle furnace at 550 deg.C for 15 minutes; cool in desiccator.
2. Weigh filter paper to nearest mg. (tare weight)
3. Insert filter into Buchner funnel. Fit in with filter paper folded neatly up sides of funnel; wet down with distilled water, using vacuum.
4. Filter a known volume of solids (largest volume that will not blind the filter), applying vacuum.
5. Carefully remove filter from funnel; dry in oven at 103 deg.C for 1 hour; cool in desiccator; weigh again (dry weight).
6. If volatile solids weight is desired, burn in muffle furnace at 550 deg.C for 20 minutes. Cool in desiccator. Weigh again (ash weight).
7. *Calculate:*

$$\text{TSS (mg/L)} = \frac{\text{dry weight(mg) - tare weight(mg)}}{\text{ml. sample}} \times 1000$$

$$\text{Fixed Solids (mg/L)} = \frac{\text{ashed weight(mg) - tare weight(mg)}}{\text{ml. sample}} \times 1000$$

$$\text{Volatile Solids (mg/L)} = \text{TSS (mg/L) - Fixed Solids (mg/L)}$$

$$\%\text{ Volatile Solids} = \frac{\text{Volatile Solids (mg/L)}}{\text{TSS (mg/L)}} \times 100$$

Total Solids/Volatile Solids

Total Solids tests are done on concentrated wastewater sludges. The result will include both suspended and dissolved solids, and are usually registered as percentages. Total solids tests are not done on process wastewaters because a large portion of the dissolved solids in wastewater samples may be inorganics originating from the carriage water, and not from the waste. In sludges, however, only a small portion of the solid weight is dissolved; almost all of it is suspended. Since filtration of this type of sample is extremely difficult, a suspended solids test is not practical, and the Total Solids test is performed.

Treatment Plant Significance

Anaerobic Digester Sludge: Total Solids test on Primary Sludge (digester influent) yields data on status of primary clarification and influent sludge pumping. Its concentration (plus Volatile Solids percentage) will affect performance of anaerobic digester and, in turn, dewatering and drying units. With this and digested sludge Total Solids content, digester efficiency and % Reduction Volatile Solids can be calculated.

Total Solids data is used to calculate supernatant and dewatering filtrate return flows, and sludge pumping rates. Percent Volatile Solids in sludge refers to efficiency of pretreatment and chemical characteristics of plant influent wastewater. Nonvolatiles in the digester are not usable by the organisms, and reduce effective volume of the digester.

Aerobic Digester Sludge: The Total Solids tests have similar significance, but refer to condition of process in the secondary clarifier. Total Solids tests are also performed on digester supernatant.

Expected TS Primary Sludge: 3-5%
Expected TS Anaerobically Digested Sludge: 7-10%
Expected TS Dewatered Sludge (An.Dig.): 30-40%
Expected TS Aerobically Digested Sludge: 2-5%
Expected TS Dewatered Sludge (Aer.Dig.): 6-9%

Treatment Plant Control

Problem: Lower then normal Total Solids concentration in Primary Sludge

Possible Cause	Test This Sample Also For	Other Checks
Denitrification (rising sludge)	VS	Prim.Eff. TSS, DO Sludge pumping frequency
Sludge Coning		Prim.Eff. TSS, DO Rising Sludge TS conc.Prim. Sludge at start of pumping cycle Sludge blanket depth
Hydraulic Overload	VS	Plant Flow Prim.Eff. TSS, COD, DO
Mechanical Malfunction		Operation of sludge pump Operation of sludge collector mechanism
Incorrect Analysis		QC Procedures Refer to Standard Methods

Problem: Lower than normal Total Solids concentration in Anaerobically digested sludge.

Possible Cause	Test This Sample Also For	Other Checks
Incomplete digestion	VS	VA/A ratio Gas production, % CO_2 TSS supernatant Digester temperature, loading, mixing
Prim.sludge concentration too low	VS	Prim. sludge TS Plant Flow Prim. sludge pumping frequency
Industrial toxic shock	VS	Supernatant COD Raw wastewater COD Past plant operating records
Incorrect Analysis		QC procedures Refer to Standard Methods

Problem: Lower than normal Total Solids concentration in Secondary
Sludge

Possible Cause	Test This Sample Also For	Other Checks
Denitrification (rising sludge)	microscopic exam	TSS final effluent Sludge blanket level RAS rate Settleability
High influent BOD load	VS	Raw & prim.eff. COD, TSS, NH$_3$-N, Tot. Phos. F/M ratio, MCRT, Settleability, Final eff. TSS, COD, NH$_3$-N, Tot.Phos.
Hydraulic overload		Plant flow Raw & prim.eff COD, TSS Settleability, RAS rate
Industrial toxic shock	pH microscopic exam	Settleability TSS final effl. Contact contributing industries
Incorrect Analysis		QC procedures Refer to Standard Methods

Analysis: Total Solids/Volatile Solids WasteWater Sludge

In Total Solids test, known volume is not taken. Instead, an evaporating dish
is filled, wet weight, dry weight, ash weight is recorded. Percent Total Solids can
then be determined.

Quality Control:
* If Total Solids is being performed on a water sample, rather than a
  sludge sample, dry first at 98 degrees Celsius till the visible water is out,
  to prevent spattering, then turn up to 103 degrees Celsius.
* See Quality Control in Total Suspended Solids section.

Apparatus:
* drying oven
* muffle furnace
* desiccator
* tongs
* evaporating dish

Procedure:
1. Burn clean evaporating dish in muffle furnace for 15 minutes. Cool. Weigh dish to the nearest ten milligrams (tare weight).
2. Fill dish about 3/4 with sludge. Weigh again (wet weight).
3. Dry sludge in 103 deg.C oven overnight. Cool in desiccator. Weigh again (dry weight).
4. Burn in muffle furnace at 550 deg.C. Cool in desiccator. Weigh again (ashed weight).
5. *Calculate:*

$$\text{Total Solids (\%)} = \frac{\text{dry weight(g.) - tare weight(g.)}}{\text{wet weight (g.) - tare weight (g.)}} \times 100$$

$$\text{Total Solids (mg/L)} = \text{Total Solids (\%)} \times 10{,}000$$

$$\text{Fixed Solids (\%)} = \frac{\text{ashed weight(g.) - tare weight(g.)}}{\text{dry weight (g.) - tare weight (g.)}} \times 100$$

$$\text{Volatile Solids (\%)} = 100\% - \text{Fixed Solids (\%)}$$

$$\text{Volatile Solids (mg/L)} = \text{Volatile Solids (\%)} \times 10{,}000$$

Settleable Solids

Settleability is a quick and easy inplant control test. The ability to settle solids from wastewater has been the basis of treatment since the earliest days of pollution control, and results of this test are still important at Activated Sludge treatment plants, where the settling capacity of the secondary sludge is variable.

Treatment Plant Significance

The first settleability tests were performed on primary sludge, and results measured in ml/L. From this a calculation of sludge pumping rate was made. In more recent years, calculations of primary sludge pumping rates are made based on an expected volume and Total Solids concentration of the sludge. Primary sludges are generally easy to settle, and most often cause problems if pumping frequency is not adequate.

Secondary sludge settleability tests are much more important today, particularly for activated sludge, which can be difficult to settle. The Mallory Settleometer is designed for this use, and has a beaker-like shape (very much like a clarifier), with calibrations to measure the sludge level. In many treatment facilities it has been replaced with the one liter graduated cylinder. Most often performed on Mixed Liquor, a 30 minute settling test yields data on the performance of the secondary clarifier, and it relates to the waste/microorganism balance in the aeration tank (F/M ratio). This along with the TSS test results helps to regulate the RAS rate.

The Settling Test results can also project expectations for downstream sludge handling processes; the same test can be performed on Activated Sludge, and thickened or digested secondary sludges.

Expected Settled Volume (Activated Sludge): 200-400 ml in 1 Liter cylinder.
Expected SVI: near 100
Expected SDI: near 1

Treatment Plant Control

Problem: Mixed Liquor settling too fast; pinpoint floc in final effluent.

Possible Cause	Test This Sample Also For	Other Checks
Operational control problem	DO, MLTSS/VS	Raw prim.eff, final eff.COD, TSS, Tot.phos. F/M, MCRT, RAS rate, conc.sludge

Problem: Mixed Liquor settling too slowly.

Possible Cause	Test This Sample Also For	Other Checks
Operational control problem	DO, MLTSS/VS	Raw, prim.eff, final eff. COD, TSS, Tot.phos. F/M, MCRT, RAS rate, conc.sludge Sec.sludge blanket level
Industrial toxic shock	DO, pH, microscopic examination	raw pH sludge conc., Final eff. pH, TSS/VS, COD, Tot.phos.
Hydraulic overload	DO, MLTSS/VS	Influent flow, sludge conc., Prim.eff. TSS/VS, COD, Tot.phos. Sludge conc., final eff. TSS/VS, COD, Tot.phos.
Predominance of filamentous organisms	DO, microscopic examination	Visual-bulking sludge Final eff., prim eff., TSS/VS, COD, Tot. Phos., NH_3-N, Sec. sludge blanket level. F/M Ratio, MCRT, RAS rate, TS.sec.sludge

Analysis: Settleable Solids - Mixed Liquor

Calculations of Sludge Volume Index and Sludge Density Index for treatment process control are done based on the settled volume of sludge, and approximate calculations for activated sludge concentration can be done.

Quality Control:
- Collect sample and do test immediately or sludge will rise.
- Be sure sample has been well mixed before pouring into cylinder.
- Solids which rise to the top surface are not disturbed, and are not to be considered settleable solids.
- Keep temperature constant.

Apparatus:
- settleometer or 1 L graduated cylinder

Procedure:
1. Fill vessel to the total volume with mixed liquor as soon as possible after taking sample.
2. Allow solids to settle for 30 minutes. Mark sludge level. Graph settling rate.

6

Dissolved Oxygen

Dissolved oxygen (DO) is one of the most important and useful water measurements. Though the oxygen concentration in air is about 21%, in water it is only slightly soluble. Oxygen saturation ranges from 7 ppm in hot water to 15 ppm in cold water and is 9.2 ppm at 20 degrees Celsius and atmospheric pressure at sea level. Type and amount of biological activity in a water body will depend upon the amount of dissolved oxygen present. Most microorganisms use free or dissolved oxygen for respiration. Oxygen depletion in natural water bodies caused by the addition of bacterial nutrients (wastewater or agricultural runoff) may limit life in that water. The ability of the stream's microbes to degrade any added nutrients will be limited by the amount of dissolved oxygen in the water.

On the other hand, photosynthesis adds dissolved oxygen to the water as a waste by-product.

Treatment Plant Significance

The measurement of Dissolved Oxygen is the basis for the BOD test, and for oxygen uptake rate tests. Oxygen depletion in long, sluggish wastewater collection systems may lead to the formation of toxic atmospheres. Nitrogen, hydrogen sulfide, methane, are emitted because of anaerobic conditions in the water. Corrosion of concrete and metal surfaces is enhanced; depletion of oxygen in primary sludges will start anaerobic action, and lead to rising sludge. Oxygen solubility determines the rate at which oxygen will be absorbed by organisms in biological aeration, making it an important factor in the cost of aeration.

Oxygen concentration determines whether biological degradation of a waste will be aerobic or anaerobic. Decreased amounts of oxygen in activated sludge aeration tanks promotes the growth of filamentous organisms. A high concentration of oxygen in aeration may promote excessive rapid growth of bacteria and hinder settling of activated sludges.

Aerobic digestion processes depend upon a positive content of dissolved oxygen for good water/solids separation and nuisance free treatment. Oxygen depletion in sludge thickening and dewatering units leads to inefficient treatment, and corrosive and hazardous atmospheres.

A moderate amount of dissolved oxygen must be present in wastewater secondary effluents, so as not to deplete the oxygen content in the receiving stream. NPDES permits often specify a range of concentrations within which the final effluent DO must reside; some may be more stringent than this.

Expected Dissolved Oxygen in Raw Domestic Wastewaters: 2-7 ppm
Expected Dissolved Oxygen in Wastewater Secondary Effluents: 4-7 ppm
Expected Dissolved Oxygen in Activated Sludge Aeration Tanks: 2 ppm

Treatment Plant Control

Problem: Dissolved Oxygen lower than normal in final effluent.

Possible Cause	Test This Sample Also For	Other Checks
Industrial high strength input	COD, NH_3-N, TSS/VS	IPP records Previous Operating Reports Contact contributing industries Raw COD, NH_3-N, Tot.Phos.,TSS/VS, DO Sludge pumping DO, blower use in biological unit Sec.sludge blanket level
Inplant recycle input	COD, NH_3-N, TSS/VS	Recycle flow records Recycle COD, Tot.Phos.,NH_3-N, TSS/VS, TS Plant flow records DO biological unit Sec.sludge blanket level
Inplant operational	COD, NH_3-N, TSS/VS	Coagulant feed rate, control problem conc., type Jar Test Chemical feed pump operation Prim.sludge TS conc. Prim.Eff. TSS conc. Sludge pumping frequency Mechanical: sludge pump, sludge collector Mixed liquor conc. DO, mixing in biological unit Sludge TS conc. DO in clarifier Sec.sludge blanket level

Possible Cause	Test This Sample Also For	Other Checks
(cont.)		Biomass microscopic exam. F/M ratio, MCRT, RAS rate, hydraulic loading
Incorrect analysis	Test by another method	DO meter function Refer to <u>Standard Methods</u>

Problem: Dissolved Oxygen higher than normal in final effluent.

Possible Cause	Test This Sample Also For	Other Checks
Hydraulic overload	COD, NH$_3$-N, TSS/VS,	Plant flow records Influent water temp. Raw & Prim.Eff. COD, Tot.Phos., NH$_3$-N, TSS/VS, DO Prim.sludge TS conc., blanket level Sec.sludge TS conc.
Industrial toxic shock	pH	Raw pH pH, DO, blower use in biological unit Biomass microscopic examination Settleability of Sec.sludge Floc formation
Incorrect analysis	Test by another method	DO meter function Winkler Procedure Refer to <u>Standard Methods</u>

Problem: Dissolved Oxygen lower than normal in aerobic digester.

Possible Cause	Test This Sample Also For	Other Checks
Unstable activated sludge	COD, Temp.	Prim.Eff. TSS/VS Prim.Sludge TS conc. DO in aeration RAS, WAS rate F/M, MCRT Sec.sludge blanket level DO in clarifier
Mechanical		Blower malfunction Leak in header Clogged diffusers

Analysis: Winkler Procedure for Dissolved Oxygen (Azide Modification)

Oxygen cannot be tested for directly by chemical methods. The Winkler Procedure is an indirect method which is dependent upon the fact that dissolved oxygen oxidizes manganese ions (Mn^{++}) to a tetravalent state (Mn^{++++}) under alkaline conditions; this more reactive state of manganese is capable of oxidizing iodine ions to free iodine under acid conditions. The amount of free iodine released is equivalent to the amount of oxygen originally in the sample. The iodine is then measured by reacting with standardized sodium thiosulfate, prepared in a concentration such that: 1 ml thiosulfate = 1 mg/L DO

Chemical Reactions - Winkler Procedure:

$MnSO_4 + 2KOH \longrightarrow Mn(OH)_2 + K_2SO_4$	azide reagent added; Mn reacts with KOH—forms hydroxide—white/tan floc.
$2Mn(OH)_2 + O_2 \longrightarrow 2MnO(OH)_2$	reaction proceeds; hydroxide reacts with O_2—Mn takes tetravalent state.
$MnO(OH)_2 + 2H_2SO_4 \longrightarrow Mn(SO_4)_2 + 3H_2O$	sulfuric acid added; floc dissolves.
$Mn(SO_4)_2 + 2KI \longrightarrow MnSO_4 + K_2SO_4 + I_2$	Mn picks up the KI from azide reagent—releases its tetravalent charge, and free iodine (yellow)

Starch, an absorption indicator, added at the end of the titration, forms a blue complex with the remaining iodine, and is useful for better visibility of the end point. (Starch supports bacterial growth; shelf life is 1 month at best, unless a preservative is added).

$I_2 + 2Na_2S_2O_3 \longrightarrow Na_2S_4O_6 + 2NaI$ titrating the iodine; discharges yellow color; starch added near end for end point recognition.

Quality Control:
- The Winkler procedure is best suited to clean waters. The Azide modification eliminates interference from nitrites, but dissolved organics, suspended solids, and iron will interfere with this test.
- Ideally, dissolved oxygen testing should be done onsite, at the sampling location. Care must be taken in collecting samples; turbulence will put

extra oxygen into an undersaturated sample. Samples that will not be tested immediately should be "fixed" with sulfuric acid and sodium azide, to stop bacterial action. Sample bottles should be filled to the top and sealed, so that a change in temperature will not affect the amount of oxygen in the sample.

- Starch supports bacterial growth; shelf life is 1 month at best, unless a preservative is added).
- *Copper Sulfate/Sulfamic Acid:* When taking samples from an aeration basin, or of settled activated sludge, and there will be time in transit to the lab for measurement, use this method. These added chemicals will not alter the DO, but will kill the bacteria in the sample so they cannot use up any more oxygen before the measurement is done.
- Measure DO immediately after taking sample (onsite if possible).
- Do not shake sample.
- Do not change temperature.
- Do not dilute sample.
- Do not let air in while sampling or measuring.

Apparatus:
- BOD bottles.
- buret.

Reagents:
manganous sulfate solution: (oxidized by DO to tetravalent state, Mn^{++++})
Dissolve 480 g. $MnSO_4 \cdot 4H_2O$ in distilled water; filter and dilute to 1 L.

alkali-iodide-azide reagent: (alkali is reduced to provide tetravalent anion for the manganese; iodide provides the iodine which is directly measured; azide eliminates nitrite interference). Dissolve 500 g. NaOH and 135 g. NaI in distilled water and dilute to 1 L. Dissolve 10 g. NaN_3 in 40 ml. distilled water and add to above solution. (Caution: dissolution of NaOH is exothermic and should be done in an ice bath)

starch: (titration endpoint indicator)
Add 5 g. starch to 1 L. boiling distilled water. Let stand overnight to settle. Decant supernate and use.

standard sodium thiosulfate: (the titrant)
Dissolve 6.205 g. $Na_2S_2O_3 \cdot 5H_2O$ in distilled water. Add .4 g. solid NaOH and dilute to 1 L. Must be standardized.

potassium bi-iodate, .0021M: (to standardize the sodium thiosulfate)
Dissolve 812.4 mg. $KH(IO_3)_2$ in distilled water and dilute to 1 L.

sulfuric acid, concentrated: (dissolves the tetravalent manganese floc).

Procedure:
1. Standardize the sodium thiosulfate:
   - dissolve 2 g. KI in Erlenmeyer flask with 100 ml distilled water.
   - Add a few drops concentrated H_2SO_4 and exactly 20 ml. potassium bi- iodate solution.
   - Dilute to 200 ml and titrate iodine (yellow) with sodium thiosulfate titrant.
   - When near end of titration (pale straw color) add starch (turns blue) and continue titrating endpoint (colorless).
   - Adjust sodium thiosulfate to .025M.
2. Fill 300 ml. BOD bottle with sample - to the top.
3. Add (submerged) 1 ml. manganous sulfate solution and 1 ml. alkali-iodide-azide reagent. Stopper, excluding bubbles, and mix by inverting a few times.
4. Let precipitate settle to half way down bottle; then add 1 ml. concentrated H_2SO_4. Mix. Wait till it dissolves.
5. Remove 200 ml. and titrate with sodium thiosulfate. Near end of titration (straw color) add starch and continue titration to endpoint.
6. 1 ml. .025 M $Na_2S_2O_3$ = 1 mg/L DO

Analysis: Membrane Electrode Method for Dissolved Oxygen (DO Meter)

The electrode method for measuring dissolved oxygen has widespread use because it eliminates interferences, enables the chemist to take measurements onsite, run oxygen profiles along the length of a process tank, and at different depths underwater. Online DO meters are often permanently installed at activated sludge aeration tanks, for onsite or remote reading.

The instrument is a milliammeter with an electrode probe on a cord. An oxygen permeable membrane is stretched across the end of the probe, and holds in the sensing electrode which is submerged in an anoxic electrolyte solution. The probe fits snugly into the neck of a BOD bottle (filled with sample). A small stirrer is attached to the end of the probe, which keeps new water flowing past the membrane as it reads. Differential pressure makes oxygen molecules pass through the membrane, where they are reduced (take on electrons) at a cathode, and then flow to an anode, oxidizing it. The result is a flow of electrons from cathode to anode proportional to the oxygen passing through the membrane. The electrical signal is then converted to concentration units, and reads out on the meter. Temperature is important, and is monitored by a thermistor built into the probe.

Quality Control:
- A DO meter must be calibrated. For daily use it can be calibrated to the oxygen saturation point (wet probe in air), but it should also periodically be calibrated to the Winkler procedure results for the same sample.
- Probe membranes should be changed frequently, according to manufacturer's directions, and whenever readings are unusual. Suspended solids collecting on the membrane will interfere with the electrode operation. Newer probes have screw-on endings, which are easy to change.
- Standardize thiosulfate.
- Careful use of Mettler balance (check with standard weights).
- Read DO immediately after sampling.

Apparatus:
- Dissolved Oxygen Meter and Probe
- BOD bottle

Procedure:
1. Calibrate DO meter (either by air calibration or by Winkler procedure).
2. Fill bottle to neck with sample.
3. Insert probe into neck of bottle; do not allow any trapped air bubbles.
4. Turn on stirrer. Read DO.

Effect of Temperature on Oxygen Saturation (at 1 atm. pressure)

Degrees Celsius	mg/L DO (sat)	Degrees Celsius	mg/L DO (sat)
0	14.6	14	10.4
1	14.2	15	10.2
2	13.8	16	10.0
3	13.5	17	9.7
4	13.1	18	9.5
5	12.8	19	9.4
6	12.5	20	9.2
7	12.2	21	9.0
8	11.9	22	8.8
9	11.6	23	8.7
10	11.3	24	8.5
11	11.1	25	8.4
12	10.8		
13	10.6		

7

Biochemical Oxygen Demand (BOD5)

BOD: The Oxygen Consumed by Bacteria in Metabolizing a Waste

The BOD test is a bioassay, a procedure in which microorganisms are grown in a somewhat natural environment, and their life processes are observed. In this case, it is oxygen use that is being measured. BOD analysis is required by federal regulation at all secondary wastewater treatment plants, and is routinely done on composite samples of Raw wastewater, Primary Effluent, Final Effluent. BOD's are also done on any internally generated recycle flow (digester supernatant), and on any industrial input to the treatment process (including landfill leachate, septage, stormwater).

Treatment Plant Significance

BOD analysis is used to indicate the organic strength of the wastewater, calculate % removals, determine amount of air needed in aeration and the organic load on secondary, design treatment units, calculate F/M ratio, determine industrial surcharges and pretreatment requirements.

Because the BOD test takes five days to complete, it has little value for immediate treatment plant control. However, it is a good approximation of microbe reaction to organic content of the water, and trends toward upset can be corrected. This is an indirect measure of the strength of the wastewater. A stronger water would have more food value for bacterial growth—and more oxygen consumed in the test. The result would be a higher BOD concentration.

The choice of a five day incubation period for this analysis was a practical decision. It was found that incubating the sample for twenty days allowed enough time for metabolism of 95% of a typical municipal waste, but it was very difficult to keep enough oxygen in the bottle for that long, and results may be higher from oxygen use by bacteria which oxidize inorganics, rather than organics in the sample. The compromise was to employ a five day test, which oxidizes about 75% of the waste.

Expected BOD in Raw Domestic Wastewaters: 200-400 ppm
Expected BOD in Wastewater Secondary Effluents: <30 ppm

Treatment Plant Control

(Although immediate plant control is not possible because of the duration of the BOD test, a trend toward higher BOD's may elicit similar responses). Review of laboratory records and plant operating conditions for the previous five days is necessary.

Problem: Higher than normal BOD in final effluent

Possible Cause	Test This Sample Also For	Other Checks
Industrial high strength input	COD, Tot.Phos., NH₃-N, TSS/VS, DO	IPP records Previous Operating Reports Contact contributing industries Raw COD, Tot.Phos., NH₃-N, TSS/VS, DO Sludge pumping DO, blower use in biological unit Sec.sludge blanket level
Inplant recycle input	COD, Tot.Phos., NH₃-N, TSS/VS, DO	Recycle flow records Recycle COD, Tot.Phos., NH₃-N, TSS/VS, TS. Plant flow records. DO biological unit Sec.sludge blanket level
Hydraulic overload	COD, Tot.Phos., NH₃-N, TSS/VS, DO	Plant flow records Influent water temp. Raw & Prim.Eff. COD,Tot.Phos., NH₃-N, TSS/VS, DO Prim.sludge TS conc. Sec.sludge blanket level, TS conc.
Inplant operational control problem	COD, Tot.Phos., NH₃-N, TSS/VS, DO	Coagulant feed rate, conc., type Jar Test Chemical feed pump operation

Possible Cause	Test This Sample Also For	Other Checks
(cont.)		Prim.sludge TS conc.
		Prim.Eff. TSS conc.
		Sludge pumping frequency
		Mechanical: sludge pump, sludge collector
		Mixed liquor conc.
		DO, mixing in biological unit Sec. sludge TS
		conc., blanket level
		Biomass microscopic examination
		F/M ratio, MCRT, RAS rate, hydraulic loading
Incorrect analysis	COD, DO (Winkler Procedure)	QC procedures Refer to Standard Methods

Analysis: Biochemical Oxygen Demand (BOD5)

The dissolved oxygen may be measured with a DO meter (DO probes are built to fit neatly in the neck of the BOD bottle), or by the Winkler procedure for dissolved oxygen. (If using the Winkler procedure, two identical sets of samples must be prepared. Winkler procedure is done on one set; the other set is incubated).

Dilution: Most wastewaters require dilution of the sample in order to successfully run this test. Filling the BOD bottle with full strength wastewater would result in total oxygen depletion long before the five days was up, yielding invalid results. To calculate BOD when dilutions are made:

$$BOD = \frac{Initial\ DO - Final\ DO}{\%\ dilution} \times 100$$

The water used to make dilutions is important. Distilled water is too pure for a bioassay; salts must be added to preserve the osmotic balance of the cells.

Routinely, for domestic wastewater, 2-4 ml of Raw wastewater is diluted to 300 ml in the BOD bottle, 10-15 ml. primary effluent is used, and 80-100 ml of final effluent is used. However, this is variable.

Seed: Clean waters, and those which have had bacteria settled from them, chlorinated wastewaters, and some industrial wastewaters, may not have enough bacteria present in the water to process this test effectively, to provide a true indication of the strength of the wastewater. Bacteria must be added (seed). As a source of seed, domestic raw wastewaters or primary effluents are often used; they contain an abundance of a wide variety of microorganisms.

However, when considering the BOD test results on a sample to which seed has been added, the oxygen depletion in that bottle has resulted from both the sample and the seed. Therefore, the BOD value of the seed must be subtracted from the result.

The most accepted way to do this is to calculate the oxygen depletion caused by the added seed, and subtract it from the oxygen depletion caused by the "sample + seed". Then complete the calculation, and BOD results will represent the sample only.

Carbonaceous BOD/Nitrogenous BOD: A good sample will contain an abundance of aerobic and facultative heterotrophic bacteria which utilize organic compounds for food, breaking them down with the use of oxygen, and providing an accurate measurement of BOD. The sample may also contain some aerobic nitrifying bacteria, which use carbon dioxide for food, but which use oxygen to oxidize ammonia to nitrate (Nitrosomonas and Nitrobacter). This oxygen use has nothing to do with the organic strength of the wastewater, but will become incorporated into the BOD test result. If desired, this Nitrogenous BOD can be eliminated by the addition of a Nitrification Inhibitor at the start of the test. The test is then referred to as CBOD.

Quality Control:
- BOD test must be performed within 24 hours of sample collection; (storage at 4 degrees Celsius).
- Calibrated thermometer should be permanently installed in BOD incubator, and checked periodically against another calibrated thermometer.
- Calibrate DO meter with each use.
- Maintain dilution water quality and seed quality. Keep dilution water and distilled water containers clean and out of direct sunlight; periodically acid wash.
- To check precision and accuracy in the operation of this test procedure, and to verify seed and dilution water quality, periodically run a standard along with the samples, using the Glucose/Glutamic Acid Method:
  1. Prepare a 300 mg/L solution of the standard (150 mg. glucose and 150 mg. glutamic acid dissolved in a liter of water).
  2. Add 6 ml. of this to a BOD bottle (2% dilution) to provide an acceptable DO depletion after incubation; add seed; fill with dilution water.
  3. Record DO; incubate for 5 days with the other samples; record DO; calculate BOD. This standard should have a BOD of 198 mg/L.

Apparatus:
- 300 ml. BOD incubation bottles with ground glass stoppers
- water seal caps
- incubator (20 deg.C)

- dilution water bottle with tube
- dissolved oxygen meter

Reagents:

phosphate buffer solution: (dilution water)
Dissolve 8.5 g KH_2PO_4, 21.75 g. K_2HPO_4, 33.4 g. $Na_2HPO_4 \bullet 7H_2O$, and 1.7 g. NH_4Cl in 500 ml. distilled water. Dilute to 1 L. Prepare weekly; store in refrigerator at 4 deg.C.

magnesium sulfate solution: (dilution water)
Dissolve 22.5 g. $MgSO_4 \bullet 7H_2O$ in distilled water. Dilute to 1 L.

calcium chloride solution: (dilution water)
Dissolve 27.5 g. $CaCl_2$ in distilled water. Dilute to 1 L.

ferric chloride solution: (dilution water)
Dissolve .25 g. $FeCl_3 \bullet 6H_2O$ in distilled water. Dilute to 1 L.

sodium sulfite solution: (dechlorinator)
Dissolve 1.575 g. Na_2SO_3 in water. Dilute to 1 L. Prepare fresh daily.

glucose/glutamic acid solution: (to check dilution water quality)
Dry glucose and glutamic acid at 103 deg.C for 1 hr. Add 150 mg each of these to distilled water. Dilute to 1 L. Prepare fresh for each use.

nitrification inhibitor: purchased.

seed water: (to provide microorganisms for bacteria deficient samples).
Obtain primary or final effluent from wastewater treatment plant. Add seed to dilution water or directly to each sample after dilution.

Procedure:

1. Prepare dilution water:
   - To each liter of distilled water add 1 ml. each of phosphate buffer, $MgSO_4$, $CaCl_2$, $FeCl_3$. Mix well.
   - Check dilution water for presence of toxics periodically by running a bottle of glucose/glutamic acid solution with a 2% dilution. Add seed. Results should indicate a BOD of 198 mg/L.
2. Adjust sample pH to 6.5-7.5.
3. If sample contains chlorine, dechlorinate with Na_2SO_3.
4. Bring sample to 20 deg. C before making dilutions.
5. If sample is cold or over-aerated, agitate to liberate supersaturated oxygen.

6. If DO of sample is less than 7 ppm, shake to add oxygen; let it sit for a few minutes; check DO again, and proceed.

7. If CBOD results are desired, add 3 mg. nitrification inhibitor to diluted sample in BOD bottle, and 10 mg/L nitrification inhibitor to dilution water.

8. Dilute sample with dilution water so that after 5 day incubation sample will be expected to have at least 1 mg/L DO, but it will have used up at least 2 mg/L of the DO present at initial reading. Depletions beyond this range yield an invalid test.

9. If sample requires seeding, add 2-3 ml. seed to each BOD bottle, or to dilution water (5-10 ml per liter).

10. Run a sample of properly diluted seed water to obtain DO depletion/ml. seed.

11. Run a blank to check purity of dilution water. Fill a BOD bottle with unseeded dilution water. If DO depletion after incubation is more than .2 mg/L the test is considered invalid; correct future quality of dilution water.

12. With calibrated DO meter, read initial DO on sample, seed water, glucose/glutamic acid check, and blank. Stopper bottle, (no air bubbles). Add water seal and seal cap.

13. Incubate for 5 days at 20 deg. C.

14. With calibrated DO meter, read final DO on all bottles.

15. Calculate BOD. Subtract DO depletion for seed. Make correction for any excess DO depletion in blank bottle.

Chemical Oxygen Demand

This test is a good estimate of the oxygen demand for most domestic wastewaters, and it is used extensively for analysis of industrial wastes. It is not preferred by EPA, however, and BOD is still the required test at municipal and most industrial wastewater treatment plants. COD is often performed along with the required BOD; it is a measure of all the oxidizable organics in the water. The COD test chemically oxidizes almost all organics which can be oxidized (some hydrocarbons and benzyl derivatives are not oxidized, but most fats, proteins, carbohydrates and celluloses are).

Treatment Plant Significance

COD analysis is used to indicate the organic strength of the wastewater, calculate % removals, determine amount of air needed in aeration and the organic load on secondary, design treatment units, determine industrial surcharges and pretreatment requirements.

COD oxidized compounds include nonbiodegradables, which a BOD can't measure, and toxics, which will inhibit the BOD test. However, it does not measure how rapidly the waste would be biologically degraded in the receiving water, and it should be considered an independent measure of wastewater strength, and not the same as a BOD test. The greatest advantage of the COD test is that it can be performed in a couple of hours, whereas a BOD takes five days.

Expected COD in Raw Domestic Wastewaters: 300-600 ppm
Expected COD in Wastewater Secondary Effluents: 30-80 ppm

Treatment Plant Control

Problem: Higher than normal COD in final effluent

Possible Cause	Test This Sample Also For	Other Checks
Industrial high strength input	Tot.Phos.,NH$_3$-N, TSS/VS, DO	IPP records Previous Operating Reports Contact contributing industries Raw COD, Tot.Phos.,NH$_3$-N, TSS/VS, DO, Sludge pumping DO, blower use in biological unit Sludge blanket level
Industrial toxic	NH$_3$-N, TSS/VS, DO	Plant operating records Contact contributing industries Raw pH DO, blower use in biological unit Biomass microscopic examination Floc formation Settleability, TS conc. sec.sludge Sludge blanket level
Inplant recycle input	Tot.Phos., NH$_3$-N, TSS/VS, DO	Recycle flow records Recycle COD, Tot.Phos., NH$_3$-N, TSS/VS, TS. Plant flow records DO biological unit Sec.sludge blanket level
Hydraulic overload	Tot.Phos., NH$_3$-N, TSS/VS, DO	Plant flow records Influent water temp. Raw & Prim.Eff. COD, Tot.Phos., NH$_3$-N, TSS/VS, DO Prim.sludge TS conc. Sludge blanket level, TS conc.
Inplant operational control problem	Tot.Phos., NH$_3$-N, DO TSS/VS	Coagulant feed rate, conc., type; Jar Test Chemical feed pump operation Prim.sludge TS conc. Prim.Eff. TSS conc. Sludge pumping frequency Mechanical: sludge pump, sludge collector Mixed liquor conc. DO, mixing in biological unit Sludge TS conc., blanket level Biomass microscopic examination F/M ratio, MCRT, RAS rate, hydraulic loading
Incorrect analysis		QC procedures Refer to <u>Standard Methods</u>

Analysis: Chemical Oxygen Demand - Open Reflux Method

In this test organic substances in the sample are oxidized with an excess of potassium dichromate (a strong chemical oxidant) under strongly acidic, hot conditions.   The unreacted dichromate is measured by titrating with ferrous ammonium sulfate, a reducing agent, and then a subtraction is done to see how much oxygen was used. The COD test yields results which are about 1½-2 times that of a BOD on domestic wastewater.

Quality Control:
*   The titrant must be standardized (FAS).
*   The standard must be dried in an oven (it picks up water on standing)

Apparatus:
*   reflux condenser
*   refluxing flask
*   buret
*   hot plate

Reagents:
standard potassium dichromate, .0417M: (the standard)
Dissolve 12.259 g. $K_2Cr_2O_7$ (dried at 103 deg.C for 2 hours) in distilled water; dilute to 1 L.

sulfuric acid reagent: (catalyzes the digestion)
Add 10 g. Ag_2SO_4 to 1 L concentrated H_2SO_4; allow to stand for 1-2 days to dissolve.

ferroin solution: (titration endpoint indicator)
purchase prepared ferroin indicator.

standard ferrous ammonium sulfate solution, .25M: (the titrant)
Dissolve 98 g. FAS in distilled water; add 20 ml. concentrated H_2SO_4; cool; dilute to 1 L. Must be standardized.

mercuric sulfate: dry (catalyzes the digestion)

Procedure:
1. Standardize the FAS:
    *   dilute 10 ml standard $K_2Cr_2O_7$ to 100 ml.
    *   add 30 ml concentrated H_2SO_4; cool.
    *   add 2-3 drops ferroin
    *   titrate with FAS to brownish red color change.

2. Place 50 ml sample in 500 ml. refluxing flask. Add 1 g. $HgSO_4$ and several boiling chips. Run a distilled water blank also.
3. Very slowly add sulfuric acid reagent, with mixing. Cool.
4. Add 25 ml. $K_2Cr_2O_7$ and mix.
5. Attach flask to condenser; turn on cooling water. Add remaining sulfuric acid reagent through open end of condenser, mixing.
6. Reflux for 2 hours. Cool and wash down condenser with distilled water.
7. Disconnect condenser; dilute mixture to twice its volume; cool to room temperature.
8. Titrate to reddish brown endpoint with FAS to measure unreacted dichromate, using 2-3 drops ferroin indicator.
9. *Calculate COD:*

$$COD \ (mg/L) = \frac{(A - B) \times molarity \ FAS \times 8000}{ml \ sample}$$

where: A=ml.FAS used for blank
         B=ml.FAS used for sample

Phosphorus

Phosphorus is a vital nutrient for all living things. Cellular phosphate compounds trap energy generated from food consumed and transfer it to activities which demand it: locomotion, reproduction, growth. Without the phosphorus to build these energy compounds, cell life cannot exist.

However, excessive phosphorus in natural water bodies stimulates bacterial and algal growth. Massive blooms may occur with resultant deposition of excessive solids which eventually fill up the water body, shortening its life. This process, called eutrophication, is a natural process, but one which should occur over hundreds or thousands of years. Excessive nutrients discharged into our lakes had hastened the solids accumulation to the point of threatening the existence of even the Great Lakes.

Phosphorus occurs naturally almost solely as phosphates. Most phosphates are dissolved, but some are in combination with suspended particles in the water. About 3 mg/L of inorganic phosphorus occurs in domestic wastewater from the breakdown of protein wastes. Most of the rest comes from synthetic detergents and industrial cleaning preparations. Biological wastewater treatment processes can only remove about 2 ppm phosphorus (bacteria will only take what they need to sustain life), and the rest passes to the receiving stream. In the late 1970's, many individual state modifications to the Clean Water Act required that treatment plants, especially those discharging to the Great Lakes, limit the amount of phosphorus discharged to receiving waters.

Other contributions to the phosphorus content of wastewaters are from fertilizer runoff and from polyphosphates added to control corrosion or scaling in water distribution systems and in boiler feedwater.

Phosphates occur in three forms:

Orthophosphate: Simple phosphates, or reactive phosphate; Ex. Na_3PO_4 sodium phosphate (tribasic), NaH_2PO_4 sodium phosphate (monobasic). Orthophosphate is the only form of phosphate that can be directly tested for in the laboratory; it is the form that bacteria use directly for metabolic processes.

Polyphosphate: Acid hydrolyzable phosphate; Ex. $Na(PO_3)_x$ sodium hexametaphosphate (Calgon). Polyphosphates come from detergents and water additives. They can be converted to orthophosphate by acid addition and boiling of the sample.

Organic Phosphate: Mostly from industrial process sources; in most municipal wastewaters, this is a small amount. It can be converted to orthophosphate by digestion with an oxidizing agent under strong acid conditions. This digestion also converts polyphosphate to orthophosphate. All phosphates will gradually hydrolyze in natural waters to the ortho form.

Total Phosphorus is the total amount of phosphorus in the sample after all forms have been converted to orthophosphate. It is Total Phosphate which is regulated in wastewater treatment plant effluents, and which is most often tested for.

Treatment Plant Significance

Phosphorus removal in wastewater treatment is most often achieved by the addition of coagulant metal salts, frequently combined with an anionic polymer. The coagulants react with water to produce floc particles which adsorb suspended solids in the water, and settle them out. In this manner, phosphorus components which are associated with solid particles are removed, and become part of the sludge. Some dissolved phosphorus will also be converted to particulate form, and be removed. Not all treatment plants need to add chemicals to remove phosphorus. If the bacteria treating the waste will remove enough phosphorus to effect compliance with the NPDES permit, no chemical treatment may be needed.

Nutrient Deficiency: Many industrial wastewater treatment plants are treating a water which is phosphorus deficient. If biological treatment is employed, the bacteria which treat the waste must have enough phosphorus to sustain life. Phosphorus must be added, either as phosphoric acid or a high phosphate fertilizer.

Expected Total Phosphorus in Raw Domestic Wastewaters: 3-10 ppm
Expected Total Phosphorus in Wastewater Secondary Effluents: .1-7 ppm

Treatment Plant Control

Problem: Higher than normal Total Phosphorus in final effluent

Possible Cause	Test This Sample Also For	Other Checks
Industrial high phosphorus input	COD, NH_3-N, TSS/VS	IPP records Previous Operating Reports Contact contributing industries Raw COD, NH_3-N, Tot.Phos., TSS/VS, DO Primary sludge pumping DO, blower use in biological unit Sec.sludge blanket level
Industrial toxic shock	COD, NH_3-N, TSS/VS, DO	Plant operating records Contact contributing industries Raw pH DO, blower use in biological unit Biomass microscopic examination Floc formation Settleability, sec.sludge Sludge TS conc., blanket level
Inplant recycle input	COD, NH_3-N, TSS/VS, DO	Recycle flow records Recycle COD, Tot.Phos., NH3-N, TSS/VS, TS Plant flow records DO biological unit DO in sec. clarifier Sludge blanket level
Hydraulic overload	COD, NH_3-N, TSS/VS, DO	Plant flow records Influent water temp. Raw & Prim.Eff. COD, Tot.Phos., NH_3-N, TSS/VS, DO Prim. sludge TS conc. Sludge blanket level Sludge TS conc.
Inplant operational Control Problem	COD, NH_3-N, TSS/VS, DO	Coagulant feed rate, conc., type Jar Test Chemical feed pump operation Prim.sludge TS conc. Prim.Eff. TSS conc. Primary sludge pumping frequency Mechanical: sludge pump, sludge collector Mixed liquor conc.

	DO, mixing in biological unit
	Sludge TS conc.
	DO in clarifier
	Sec.sludge blanket level
	Biomass microscopic examination
	F/M ratio, MCRT, RAS rate, hydraulic loading
Incorrect analysis	QC procedures
	Refer to <u>Standard Methods</u>

Analysis: Total Phosphorus—Persulfate Digestion Method for Conversion of Organic Phosphorus and Polyphosphorus to Orthophosphorus

Persulfate Digestion is a safe, simple procedure for conversion of more complex forms of phosphorus to orthophosphate.

Quality Control:
- Glassware for phosphorus testing must be acid washed to eliminate any soap residue which may contain phosphorus.
  Acid Washing: Wash in hot water with phosphorus free soap; rinse twice with 10% HCl solution, then with distilled or deionized water.

Wastewater Treatment plant chemists usually keep a separate set of glassware (periodically acid washed) to be used for phosphorus testing only.

- Samples for phosphorus testing should not be kept more than 2 days.
- Samples should be acidified and kept refrigerated at 4 degrees Celsius if they must be stored at all.

Apparatus:
- acid washed glassware
- hot plate

Reagents:
<u>sulfuric acid solution</u>: (catalyzes the digestion)
Add 300 ml concentrated sulfuric acid to 600 ml distilled water; dilute to 1L.

<u>sodium hydroxide, 1N</u>: (to neutralize sample after digestion)
Dissolve 40 g. NaOH in distilled water; dilute to 1L.

<u>ammonium persulfate</u>: (digestion reagent)
$(NH_4)_2S_2O_8$; dry, solid.

Procedure:
1. Adjust sample to phenolphthalien end point.
2. To 50 ml. sample add 1 ml. sulfuric acid solution and .4 g ammonium persulfate.
3. Boil gently for 40 minutes, or until volume is down to 10 ml.
4. Cool. Dilute to 30 ml with distilled water. Neutralize to phenolphthalien end point with NaOH.
5. Make up to 50 ml with distilled water. (If precipitate forms, ignore it. It will redissolve in the acid conditions of the orthophosphorus test).

Analysis: Total Phosphorus—Ascorbic Acid Method for Determining Orthophosphate

Ammonium Molybdate and Potassium Antimonyl Tartrate react under acid conditions with orthophosphate to produce a blue phospho-molybdic acid complex. Compared to orthophosphate standards, and graphed, concentration can be determined. Absorbance is read on the spectrophotometer at a wavelength of 880 nm., out of the visible light range. Older spectrophotometers which cannot achieve the higher wavelengths may find a good peak at 650 nm. (also accepted by EPA). The Ascorbic Acid method is the most accurate method for waters with small concentrations of phosphorus (under 1ppm). When testing raw wastewater by this method, dilute the sample to fall within the standards.

Quality Control:
- Phosphate standards must be weighed out carefully, and exactly.
- Standards should be carried through the entire digestion procedure along with the samples.
- Allow the same amount of time for color development for all standards and samples.
- Compare sample concentration calculated from slope against sample concentration visually observed on graph.

Apparatus:
- acid washed glassware
- spectrophotometer

Reagents:
sulfuric acid, 5N: (catalyzes reaction)
Dilute 70 ml. concentrated H_2SO_4 to 500 ml. with distilled water.

potassium antimonyl tartrate solution:
$K(SBO)C_4H_4O_6 \bullet 1/2H_2O$; (reacts to form complex which can be reduced by ascorbic acid to produce blue color in presence of phosphorus. Dissolve

1.3715 g. potassium antimonyl tartrate in 400 ml. distilled water; dilute to 500 ml.

ammonium molybdate solution: $(NH_4)_6Mo_7O_{24} \cdot 4H_2O$; (reacts to form complex which can be reduced by ascorbic acid to produce blue color in presence of phosphorus). Dissolve 20 g. ammonium molybdate in 500 ml distilled water.

ascorbic acid, .1M: (reduces the above reagents to produce blue color in the presence of phosphorus)
Dissolve 1.76 g. ascorbic acid in 100 ml. distilled water.
Stable for 1 week.

combined reagent: (produces the color)
Mix, in this order, 50 ml. 5N H_2SO_4, 5 ml. potassium antimonyl tartrate solution, 15 ml. ammonium molybdate solution, and 30 ml ascorbic acid solution. Mix well after addition of each. Stable for 4 hours.

stock phosphate solution: (the standard)
Dissolve 219.5 mg. anhydrous KH_2PO_4 in distilled water and dilute to 1 Liter.

Procedure:
1. Acid wash all glassware with hot dilute HCl.
2. If turbidity exists in raw sample, filter through .45 um membrane filter. If it is present in digested sample, ignore it. Acidic reagent should dissolve it.
3. If expected sample concentration is above 2 mg/L, dilute sample.
4. Adjust sample pH to phenolphthalien end point.
5. Prepare working standards to cover range.
6. To standards, reagent blank and sample add 8 ml combined reagent. Mix. Wait 10 min for color development.
7. Read absorbance on spectrophotometer (880 nm). Prepare calibration curve and calculate sample concentration.

10

Nitrogen

The compounds of nitrogen are of great interest to the wastewater chemist because of the importance of nitrogen in life's processes. The chemistry of nitrogen is complex; this element can take seven oxidation states; five of these are of direct interest: Nitrogen gas (N_2), organic nitrogen (C-C-C-NH_2), ammonia (NH_3), nitrite (NO_2^-), and nitrate (NO_3^-). The atmosphere is earth's source of nitrogen, where it exists as the elemental gas and is constantly removed and recycled for life's needs.

Electrical discharge and nitrogen fixing bacteria make atmospheric nitrogen available to plants for growth; animals in turn consume and break down plant protein for their own needs. The decomposition of plant and animal matter produces ammonia, which is oxidized by nitrifying bacteria (Nitrosomonas & Nitrobacter) to nitrate for plant use.

Synthetic fertilizers contain ammonium and nitrate compounds; the soil does not hold these materials well, and nitrates applied in excess of plant needs percolate to groundwater. Under anaerobic conditions, nitrates are reduced to nitrogen gas, which escapes back to atmosphere.

Before the BOD and Coliform tests were developed, chemists used tests for nitrogen to determine the sanitary quality of waters, and the strength of wastewaters. Freshly polluted waters have high levels of ammonia, which gradually are oxidized by bacteria to nitrite, then to nitrate. Where wastewater sludges are applied to land, both ammonia and nitrate are monitored to determine crop uptake and prevent leaching to groundwater.

In environmental testing, nitrogen is split into three categories:

Organic Nitrogen —   the sum of the concentration of these two is called
Ammonia Nitrogen —   <u>Total Kjeldahl Nitrogen</u>.

Inorganic Nitrogen—Nitrites and Nitrates

Nitrogen concentrations are designated in terms of the weight of nitrogen atoms in the sample, and the concentration is stated this way, regardless of which form of nitrogen is being considered.

Example: 10 mg/L NO_3, as N   or   10 mg/L NO_3-N

This means that the ion being measured is nitrate, but every liter of this solution has 10 mg. of N in it. The weight of the oxygen isn't being counted.

Organic nitrogen

Organic nitrogen is the sum of all the nitrogen present in organic compounds: amines, amides, amino acids, nitro derivatives, and others. The test procedure which digests organic nitrogen is the initial step of the Total Kjeldahl Nitrogen test.

Treatment Plant Significance

Organic nitrogen usually occurs in only small amounts in domestic wastewater. It's domestic source is urea, a major compound of urine, most of which is converted to free ammonia in wastewater collection systems. Industrial wastewaters may contribute significant amounts of organic nitrogen (protein wastes from food processing, and synthetics from the fabric industry). The test for Total Kjeldahl Nitrogen, or the test for Organic Nitrogen may be required for industrial waste samples.

Analysis: Macro-Kjeldahl Digestion Method for Organic Nitrogen

Samples are digested with concentrated sulfuric acid and a catalyst to break up the organic compounds, releasing the nitrogen as ammonia. The procedure is done with exhaust apparatus to vent toxic sulfite fumes which are produced. The sample is then transferred to a distillation flask, and the ammonia is distilled off. A titration or colorimetric procedure is performed to measure the resulting total free ammonia concentration (Total Kjeldahl Nitrogen).

The organic nitrogen concentration is the difference between the Total Kjeldahl Nitrogen result, and the result of a test for free ammonia only, in the original sample.

Quality Control:
• If colorimetric measurement is to be done, run standards and blank through digestion along with sample.

- An excess of 10 mg/L nitrate, a large quantity of dissolved salt, inorganic solids, or organics may interfere with the digestion. To avoid this, be sure that there is at least 1 ml sulfuric acid added per gram of salt in the sample (or 1 ml sulfuric acid per every three grams COD in the sample).

Apparatus:
- 800 ml Kjeldahl flasks
- heating unit capable of 370 deg.C with fume ejection apparatus

Reagents:
mercuric sulfate solution: (digestion catalyst)
Dissolve 8 grams red mercuric oxide in 10 ml 6N H_2SO_4.

digestion reagent: dissolve 134 grams K_2SO_4 in 650 ml. water and 200 ml concentrated H_2SO_4. Add, stirring, 25 ml $HgSO_4$ solution. Dilute to 1 L.

sodium hydroxide - sodium thiosulfate reagent: (neutralizes digestion reagent and raises pH for distillation)
Dissolve 500 grams NaOH and 25 grams $Na_2S_2O_3 \bullet 5H_2O$ in water; dilute to 1 L.

borate buffer: (neutralize strong acids in sample)
Add 88 ml .1N NaOH solution to 500 ml .025M sodium tetraborate ($Na_2B_4O_7$) solution; dilute to 1 L.

sodium tetraborate solution, .025 M: (for borate buffer)
Dissolve 9.5 grams $Na_2B_4O_7 \bullet 10H_2O$ in water; dilute to 1 L.

sodium hydroxide, 6N: (converts ammonia in sample to gaseous state)
Dissolve 240 grams NaOH in water; dilute to 1 L.

Procedure:
1. Select sample volume from the following:

Organic Nitrogen, mg/L	Sample Volume, ml.
0-1	500
1-10	250
10-20	100
20-50	50
50-100	25

2. Make sample up to 500 ml. Add 25 ml. borate buffer and then 6N NaOH till pH is 9.5. Add boiling chips and boil off 300 ml. to drive off free ammonia. Cool.
3. Carefully add 50 ml. digestion reagent. Add boiling chips; mix. Attach to ejection equipment and heat till volume is about 30 ml. and copious white fumes are observed. Digest for 30 minutes more. Cool.
4. Dilute to 300 ml. with water; mix.
5. Carefully add 50 ml hydroxide-thiosulfate reagent to form an alkaline layer at flask bottom. Connect to distillation apparatus; shake and distill off free ammonia.

Ammonia Nitrogen

Ammonia is a product of the microbiological decay of animal and plant protein. When dissolved in water at a pH of 7 or less, almost all of it occurs as the ammonium ion (NH_4^+). As the pH rises, it converts to the unionized gaseous form (NH_3). In any water, the two forms of ammonia are in equilibrium:

$$NH_4^+ \rightleftharpoons NH_3$$

Ammonium ion is a potential nutrient to microorganism life; upon oxidation in the receiving stream, it stimulates the growth of bacteria and algae. Unionized ammonia (the gas) is toxic to many organisms; even tiny quantities can be detrimental to life in the stream.

Treatment Plant Significance

Although organic wastes can be stabilized without carrying the biological oxidation to the nitrification stage, many wastewater treatment plants have ammonia limits stated in the NPDES permit, and nitrification will be required. It is difficult to remove ammonia from wastewater with standard process. Most of the organic carbon components in the water will be oxidized first, by carbonaceous bacteria (heterotrophs), before the nitrifiers start to work at changing ammonia to nitrate. It takes a long length process (biodisc, oxidation ditch, extended aeration), or a highly efficient process to nitrify. Treatment units may have to be expanded. Air stripping is an effective (but expensive) method for removing ammonia from potable waters and treated wastewaters, on the basis that it is a volatile component which can be scrubbed out.

Ammonia discharged puts a significant oxygen demand on the waterway. A clean stream will have many aerobic nitrifying bacteria, which will act on the discharged ammonia, converting it to nitrate, and consuming oxygen.

A source of nitrogen is necessary for bacterial metabolism in wastewater treatment. It must be added to nitrogen deficient industrial wastewaters, usually as urea or ammonium hydroxide.

When chlorine is added to any water which has ammonia present - whether it is a trace amount of ammonia in a clean potable water source, or a great deal of ammonia in a wastewater discharge - the chlorine will react with the ammonia to form chloramines, (Combined Chlorine Residual).

Expected free ammonia in Raw Domestic Wastewaters: 30-40 ppm
Expected free ammonia in Wastewater Secondary Effluents:
    Activated Sludge Process - 5-30 ppm
    Biodisc Process - 2-10 ppm

Treatment Plant Control

Problem: Higher than normal Ammonia Nitrogen in final effluent

Possible Cause	Test This Sample Also For	Other Checks
Industrial high ammonia input	COD, Tot.Phos., TSS/VS, DO	IPP records Previous Operating Reports Contact contributing industries Raw COD, Tot.Phos.,NH_3-N, TSS/VS, DO Primary sludge pumping DO, blower use in biological unit Sec.sludge blanket level
Inplant recycle input	COD, Tot.Phos., TSS/VS, DO	Recycle flow records Recycle COD, Tot.Phos.,NH_3-N, TSS/VS, TS. Plant flow records DO biological unit Sec.sludge blanket level
Hydraulic overload	COD, Tot.Phos., TSS/VS, DO	Plant flow records, Influent water temp. Raw & Prim.Eff. COD, Tot.Phos., NH_3-N,TSS/VS, DO Prim.sludge TS conc. Sludge blanket level, TS conc.
Inplant operational control problem	COD, Tot.Phos., TSS/VS, DO, pH	Coagulant feed rate, conc., type Jar Test Chemical feed pump operation Prim.sludge TS conc. Prim.Eff. TSS conc. Sludge pumping frequency Mechanical: sludge pump, sludge collector Mixed liquor conc.

Possible Cause	Test This Sample Also For	Other Checks
(cont.)		DO, mixing in biological unit
		Sludge TS conc., blanket level
		Biomass microscopic examination
		F/M ratio, MCRT, RAS rate, hydraulic loading
Incorrect analysis	Test by another method	QC procedures Refer to <u>Standard Methods</u>

Analysis: Preliminary Distillation of Free Ammonia Nitrogen

When wastewater is being tested for compliance reporting, or where ammonia concentration is more than 5ppm and colorimetric or titrimetric measurement is to be used, the sample must be distilled first. The free ammonia is collected in a solution of boric acid (as ammonium borate). This eliminates any interfering ions, color or turbidity.

Quality Control:
- When testing for nitrogen components, always use distilled, deionized water to prepare reagents and standards. These ions are often present in the sample in very small amounts, and traces of nitrogen in the distilled water may throw off results.
- Run standards and blank through distillation along with sample.

Apparatus:
- distillation apparatus
- pH meter

Reagents:
ammonia-free distilled water: Pass distilled water through deionizer column appropriate for removing ammonia; store in tightly stoppered glass container. Use for preparing reagents, standards, & for making dilutions.

borate buffer: (holds pH at 9.5; prevents release of cyanates and organic nitrogen) Add 88 ml 0.1 N NaOH solution to 500 ml 0.025 M sodium tetraborate ($Na_2B_4O_7$) and dilute to 1 L.

sodium tetraborate solution, .025 M: (for borate buffer)
Dissolve 9.5 grams $Na_2B_4O_7 \bullet 10H_2O$ in water; dilute to 1 L.

sodium sulfite: (dechlorinating agent)
Dissolve 0.9 g NaSO$_3$ in water; dilute to 1 L. Prepare fresh daily. (1 ml. neutralizes 1 mg/L chlorine).

sodium hydroxide, 6N: (raises pH to change ammonium ion to gaseous ammonia)
Dissolve 240 g. NaOH in water. Dilute to 1 L.

boric acid: (receiving solution; to collect ammonia distillate as ammonium borate)
Dissolve 20 g. boric acid (H$_3$BO$_3$) in water; dilute to 1 L.

indicating boric acid: (for titrimetric method of measurement; to indicate endpoint)
Prepare boric acid solution. Add 10 ml mixed indicator solution to this. Prepare monthly.

mixed indicator solution: Dissolve 200 mg. methyl red in 100 ml 95% ethyl alcohol. Dissolve 100 mg. methylene blue in 50 ml 95% ethyl alcohol. Combine solutions. Prepare monthly.

Procedure:
1. Prepare distillation equipment: add 500 ml water and 20 ml borate buffer to each flask; adjust pH to 9.5 with 6N NaOH solution. Add boiling chips and boil for 30 minutes to distill residual ammonia from flasks. Discard solution.
2. Dechlorinate sample with sodium sulfite.
3. Put 500 ml sample (standards and blank should also be run through distillation) in distillation flask. Add 25 ml borate buffer; adjust pH to 9.5 with 6N NaOH.
4. Put 50 ml boric acid solution (indicating boric acid solution if doing acid titration) in 500 ml Erlenmeyer flask and submerge end of distillation tube in this solution.
5. Distill standards, blank and sample until 300 ml distillate have been collected in receiving solution; Remove tube from solution; turn still off.
6. Bring standards, blank and sample back to 500 ml. with water.
7. Measure ammonia by Nesslerization, acid titration, or ion electrode method.

Analysis: Nesslerization Determination Method for Free Ammonia Nitrogen

The direct Nesslerization procedure is meant for clean waters only, and works best on samples with an ammonia concentration of less than .5 ppm. If interfering ions are present (ketones, aldehydes, alcohols, amines, calcium, magnesium), distill first.

The sample is pretreated with zinc sulfate to settle out interfering ions. Then Nessler Reagent (a strongly alkaline solution of potassium mercuric iodide) is added to the sample and to a set of ammonium standards; a yellow-brown color is developed, and comparison is made to standards by spectrophotometer.

Quality Control:
- Warm up pH meter 1/2 hour before using.
- Samples to be tested for ammonia will change upon standing. Refrigerate to inhibit bacterial growth, add sulfuric acid to pH 2, and perform analysis within 24 hours.
- Accuracy of standards is of prime importance. Correct calculation of sample concentration depends upon this.
- Use small concentration standards (1,2,3,4 ppm), or the standard curve will lose linearity. Dilute the sample, then adjust calculation for the dilution later.
- Check results of one method against results of another.
- To check the distillation, run the standards and blank through this part of the procedure along with the samples.
- Keep Nessler reagent in the dark. If it develops a precipitate, prepare new.

Apparatus:
- spectrophotometer
- pH meter

Reagents:
Nessler Reagent: (produces a yellow/brown color in reaction with ammonia nitrogen)
Dissolve 100 g. HgI_2 and 70 g. KI in a small quantity of water. Add this mixture (stirring) to a cool solution of 160 g. NaOH dissolved in 500 ml water. Dilute to 1 L. Store away from sunlight. Stable for one year.

stock ammonium solution: (the standard)
Dissolve 3.819 g. anhydrous NH_4Cl (dried, 100 deg.C) in water, and dilute to 1 L. (1000 mg/L). Make dilutions as needed for working standards from this stock. Prepare fresh standard every two weeks.

zinc sulfate solution: (pretreatment for undistilled samples; settles out interfering ions)
Dissolve 100 g. $ZnSO_4 \bullet 7H_2O$ in water; dilute to 1 L.

sodium hydroxide 6 N: (raises pH to convert ammonium ion to gaseous ammonia)
Dissolve 240 g. NaOH in water. Dilute to 1 L.

EDTA reagent: (holds any remaining Ca,Mg ions in solution to eliminate turbidity interference with Nessler reagent)
Dissolve 10 g. NaOH in 60 ml water. Dissolve 50 g. EDTA (disodium salt) into this. Warm slightly to dissolve.

Procedure:
1. If sample is undistilled, pretreat:
   * dechlorinate sample; if sample is expected to have > 5 mg/L, dilute.
   * add 1 ml $ZnSO_4$ solution to 100 ml sample; mix.
   * add 6 N NaOH to obtain a pH of 10.5. Mix. Let stand for a few minutes.
   * After floc settles, filter (discard first 25 ml filtrate) or centrifuge, saving 50 ml filtrate.
   * Add 1 drop EDTA reagent. Mix.
2. Add 2 ml Nessler reagent to sample, standards, blank. Mix. Let stand for 10-30 min for color development. Be sure that time, temp., etc. are the same for all standards, blank, and sample.
3. Read on spectrophotometer at 400-450 nm. Prepare calibration curve and calculate sample concentration.

Analysis: Acid Titration for Free Ammonia Nitrogen

This is an ammonia determination method which is used for samples which already have been distilled. The ammonia distillate is collected in indicating boric acid and titrated to the original pH of the acid. This neutralizes the borate which is tied up with the ammonia in solution, thus measuring the amount of ammonia there.

Quality Control:
* A blank and standards should be carried through the distillation and titrated.

Apparatus:
- pH meter

Reagents:
sodium carbonate, .05 N: (to standardize the sulfuric acid)
Dry 5 g. Na_2CO_3 at 250 deg C for 4 hours; cool in desiccator. Dissolve 2.5 g in water; dilute to 1 L. (.05N)

sulfuric acid, .02 N: (titrant)
Mix 2.8 ml H_2SO_4 with water; dilute to 1 L. Mix well. Titrate into 40 ml .05 N sodium carbonate solution, using pH meter, to pH 5. Boil gently 5 minutes under watch glass to remove CO_2. Cool to room temperature. Finish titrating to pH inflection point. Calculate normality of acid. Dilute to .02 N.

Procedure:
1. Titrate standard sulfuric acid into distillate, until indicator turns pale lavender (see indicating bone, under distillation reagents).
2. Calculate ammonia concentration:

$$mg/L\ NH_3\text{-}N\ =\ \frac{(A\text{-}B)\ x\ 280}{ml\ sample}$$

> *where:*   A= ml. H_2SO_4 titrated for sample
>   B= ml. H_2SO_4 titrated for blank

Analysis: Ammonia Selective Electrode Method for Free Ammonia Nitrogen

The ammonia electrode is composed of a combination pH electrode with a gas permeable membrane attached to its end to separate the sample from the electrode's internal filling solution. The sample is made basic so that all free ammonia changes into the gaseous state (pH11). Dissolved ammonia gas will pass through the membrane theoretically until the partial pressure on both sides is the same. The internal filling solution, however, is acidic enough to change the gas to ammonium ion as soon as it passes through. The pH of the alkaline ammonia ion is then sensed by the pH electrode, and converted to a millivolt reading. Pressure is maintained on the outside, and the gas continues to pass through the membrane. Ammonia standards must be used, most often 1ppm, 10ppm, 100ppm.

Quality Control:
- Distillation is advised, but not necessary if ammonia concentration is more than 5 ppm and ion selective electrode is to be used for measurement. Amines, mercury and silver interfere.
- To calibrate meter, The slope of the calibration curve (millivolts per decade of concentration) should be about 90; if it drops to near 50, change the filling solution and cap.
- After addition of NaOH, wait 3-4 minutes, at least; it takes this long for the ammonia gas to build up.
- Specific ion electrodes have a usable life of one year.
- Membrane caps should be replaced every month (with new filling solution). Erratic standard readings usually means that the membrane cap should be replaced.
- Store probe in diluted standard solution.

Apparatus:
- pH meter
- ammonia selective electrode

Reagents:
ammonia-free distilled water: Pass distilled water through deionizer column appropriate for removing ammonia; store in tightly stoppered glass container. Use for preparing reagents, standards, & for making dilutions.

stock ammonium solution: (the standard)
Dissolve 3.819 g. anhydrous NH_4Cl (dried, 100 deg.C) in water, and dilute to 1 L. (1000 mg/L). Make dilutions as needed for working standards from this stock.

sodium hydroxide 10 N: (raises pH to convert ammonium ion to gaseous ammonia)
Dissolve 400 g. NaOH in 800 ml. water; cool & dilute to 1 L.

Procedure:
1. From stock solution, prepare working standards of 100ppm, 10ppm, 1ppm.
2. Immerse probe into each standard and sample on auto-stirrer, add 1 ml NaOH, wait five minutes for conversion to take place, and read.
3. Construct standard curve on four cycle semilog graph paper; (mg/l plotted against millivolts does not produce a linear curve. Use of the log paper is an attempt to straighten the line). Set up mg/L across bottom and millivolts up the side; set the lowest standard at the x-y intersection.

Nitrite Nitrogen

The nitrite ion (NO_2^-) is most unstable, and is rarely found in significant concentrations in any water. It is the partially oxidized state which occurs briefly as ammonia is oxidized, or as nitrate is reduced, and normal concentrations are well under 1 ppm.

Treatment Plant Significance

Nitrite is extremely oxygen and chlorine demanding; it has a tendency to convert to the more oxidized form (NO_3^-), and it will scavenge oxygen from the surroundings to do so. In wastewater effluents, the major significance of nitrite is with its reactivity. A treatment process which does not nitrify will have little nitrite in the effluent; the nitrogen form will be ammonia. If the treatment process does nitrify completely, the nitrogen form in the effluent will be nitrate. If the process only partially nitrifies, the effluent may contain considerable quantities of nitrite. It is at the effluent end of the process that chlorine is added for disinfection, and a demand from a few parts per million of nitrite will have a major impact, consuming the disinfectant.

The amount of nitrification occurring in wastewater treatment process is very dependent on temperature, pH, and every chemical characteristic of the wastewater which makes the process more or less efficient. A process which achieves complete nitrification today may act differently as conditions change.

Expected Nitrite in Raw Domestic Wastewaters: < 1 ppm
Expected Nitrite in Wastewater Secondary Effluents: < 1 ppm

Treatment Plant Control

Nitrite analysis is not a routine wastewater treatment plant test. It would be performed as a possible cause of low chlorine residual concentration and high coliform bacterial counts in the final effluent. Incomplete nitrification or denitrification may be the cause.

Analysis: NED Diazotization Method for Nitrite Nitrogen

The NED Diazotization procedure for nitrite is a sensitive and reliable colorimetric method which can detect nitrite in very small concentrations. Under acid conditions, nitrite reacts with NED reagent (N-(1-naphthyl)-ethylenediamine dihydrochloride) to form a wine colored azo dye which is proportional to the nitrite concentration. Comparison can be done by colorimetry using nitrite standards.

Quality Control:
- Take care preparing standards; mix well. Concentration to be measured is very small.
- Nitrite is unstable. Nitrite standards can be kept for only one day. - Samples must be tested for nitrite immediately upon collection.
- When testing final effluents, dilute sample by 2; this test is designed to read concentrations under .5ppm.

Apparatus:
- spectrophotometer

Reagents:
nitrite-free distilled water: run distilled water through appropriate demineralizer column for removing nitrite. Use this water for all reagents, standards and dilutions.

NED reagent: N-(1-naphthyl)-ethylenediamine dihydrochloride; (produces the color)
Add 100 ml. 85% phosphoric acid and 10 g. sulfanilamide to 80 ml. distilled water. When dissolved, add 1 g. NED; mix, then dilute to 1 L. with water. Store in dark bottle. Stable for 1 month.

standard Potassium Permanganate: $KMnO_4$; .05 M; (used to standardize the nitrite stock solution).
Dissolve 8 g. $KMnO_4$ in 1 L. water. Keep in brown bottle. Age for at least 1 week. Must be standardized.

nitrite stock solution: 250 mg/L; .0179M (the standard)
Dissolve 1.232 g $NaNO_2$ in water; dilute to 1 L. Must be standardized. Prepare working standards from this solution. Make fresh daily.

sodium oxalate: $Na_2C_2O_4$, .05 N (primary standard; to standardize the permanganate) Dissolve 3.35 g $Na_2C_2O_4$ in water; dilute to 1 L.

Procedure:
1. Standardize potassium permanganate:
   - dissolve 200 mg sodium oxalate in 100 ml. water.
   - add 10 ml 1+1 H_2SO_4; heat to 95 deg.C. (keep heat above 85 deg.C through entire procedure)
   - titrate permanganate solution into this till faint pink color appears, and persists for 1 minute.

- calculate: $$\text{Molarity KMnO4} = \frac{\text{g. Na}_2\text{C}_2\text{O}_4}{\text{ml titrated x .335}}$$
- adjust to .05M
2. Standardize nitrite stock solution:
   - Add, in this order, 50 ml standard .05 M $KMnO_4$, 5 ml. concentrated H_2SO_4, and 50 ml stock nitrite solution.
   - warm to 80 deg.C on hot plate.
   - Titrate $Na_2C_2O_4$ into this solution until permanganate color is just barely discharged.
   - calculate molarity of stock solution.
3. Prepare working standards (.2-1.0 ppm, or expected range); 50 ml. of each.
4. Filter sample to remove turbidity.
5. Add 2 ml. NED reagent to standards, blank, and sample. Wait 10 min for color development.
6. Read on absorbance on spectrophotometer; 543 nm. Prepare calibration curve and calculate concentration.

Nitrate Nitrogen

Nitrate (NO_3^-) is the stable, most oxidized form of nitrogen found in natural waters. It is normally present in only small concentrations (< 1ppm). When larger concentrations occur in ground or surface waters, the most common cause is runoff or seepage from excessive agricultural fertilizer applications.
Nitrate in potable waters is limited (Primary MCL 10 ppm) because it is believed to contribute to the occurrence of methylemoglobinemia in infants (blue baby).

Treatment Plant Significance

Nitrate in wastewater discharges is usually acceptable, and in a non-nitrifying treatment process, is usually under 5 ppm. The ion is stable, does not have an oxygen or chlorine demand, and does no harm to life in the stream. Under certain conditions, however, if the wastewater effluent discharge is large compared to the volume of the receiving stream, and if downstream there is a community potable water intake, the wastewater treatment plant may be required to denitrify. When nitrate removal is necessary, an anoxic tank (detention time - no oxygen added) which breeds denitrifying anaerobes, is installed as part of the process. These organisms strip oxygen from the nitrate, reducing it to nitrogen gas, which is then expelled.   (This anaerobic condition may also strip phosphorus from the organisms, leaving it dissolved in the water, and putting the plant in noncompliance with the phosphorus limit. Denitrification process must be operated carefully).

$$\text{Oxidation} \qquad\qquad\qquad \text{Reduction}$$

$$NH_3 \text{------->} NO_2^- \text{------->} NO_3^- \text{------->} NO_2^- \text{------->} N_2$$
$$\quad\textit{Nitrosomonas} \qquad \textit{Nitrobacter} \qquad \text{Denitrifying Anaerobes}$$

$$\text{Nitrification} \qquad\qquad\qquad \text{Denitrification}$$

Expected Nitrate in Raw Wastewaters: <5 ppm
Expected Nitrate in Wastewater Secondary Effluents: 1-30 ppm (depends
   upon nitrification).

Treatment Plant Control:

Problem: Lower than normal nitrate concentration in final effluent.

Possible Cause	Test This Sample Also For	Other Checks
Incomplete nitrification	COD, DO, NH_3 Temperature	Plant Flow, COD, TSS raw wastewater, prim.eff. DO,pH biological unit, RAS rate, TSS Mixed liquor, F/M ratio, MCRT
Nitrogen deficient industrial input		NH_3, NO_3 raw wastewater Contact contributing industries
Incorrect analysis		QC procedures Refer to Standard Methods

Analysis: Cadmium Reduction Method for Nitrate Nitrogen

Nitrate is reduced to nitrite by passing sample through a column packed with activated cadmium. The sample is then measured quantitatively for nitrite.

Quality Control:
- Assure accuracy of standards.
- Compare results of this method with results of nitrate electrode method.
- Test should be done within 48 hours of sampling. The nitrate present is fairly stable, but with standing the ammonia and nitrite in the

sample will oxidize, yielding false high results. If septic conditions occur, nitrate will be reduced to nitrogen gas.
- Store Nitrate electrode in 100 ppm Nitrate standard solution. Replace endings monthly; replace electrode after one year.
- Nitrate standards have a shelf life of two weeks.

Apparatus:
- reduction column
- spectrophotometer

Reagents:
nitrate-free distilled water: Pass distilled water through deionizer column appropriate for removing nitrate; store in tightly stoppered glass container. Use for preparing reagents, standards, and for making dilutions.

copper-cadmium granules: (reduction material)
Wash 25 g. Cd granules (40-60 mesh) with 6N HCl and rinse with water. Swirl Cd with 100 ml. 2% $CuSO_4$ solution for 5 minutes. Decant and repeat until brown precipitate begins to develop. Flush with water to remove all precipitate.

NED reagent: N-(1-naphthyl)-ethylenediamine dihydrochloride; (produces the color)
Add 100 ml. 85% phosphoric acid and 10 g. sulfanilamide to 80 ml. water. When dissolved, add 1 g. NED; mix, then dilute to 1 L. with water. Store in dark bottle. Stable for 1 month.

ammonium chloride-EDTA solution: (activates reduction column)
Dissolve 13 g. NH_4Cl and 1.7 g. EDTA in 900 ml. water. Adjust to pH 8.5 with concentrated NH_4OH and dilute to 1 L.

dilute ammonium chloride-EDTA solution: (to rinse reduction column)
Dilute 300 ml. NH_4Cl-EDTA to 500 ml. with water.

copper sulfate solution, 2%: (to prepare Cd granules)
Dissolve 20 g. $CuSO_4 \cdot 5H_2O$ in 500 ml. water and dilute to 1 L.

stock nitrate solution: (the standard)
Dry potassium nitrate (KNO_3) at 105 deg C for 24 hrs. Dissolve .7218 g. in water and dilute to 1 L. (100 mg/L NO_3-N). Prepare fresh every two weeks. Use this solution to prepare working standards.

<u>stock nitrite solution</u>: 250 mg/L; .0179M (the standard)
Dissolve 1.232 g NaNO$_2$ in water; dilute to 1 L. <u>Must be standardized</u>.
Prepare working standards from this solution. Make fresh daily.

<u>standard potassium permanganate</u>: KMnO$_4$; .05 M; (used to standardize
the nitrite stock solution).
Dissolve 8 g. KMnO$_4$ in 1 L. water. Keep in brown bottle. Age for at
least 1 week. <u>Must be standardized</u>.

<u>sodium oxalate</u>: Na$_2$C$_2$O$_4$, .05 N (primary standard; to standardize the
permanganate)
Dissolve 3.35 g Na$_2$C$_2$O$_4$ in water; dilute to 1 L.

Procedure:
1. Standardize potassium permanganate:
   - dissolve 200 mg sodium oxalate in 100 ml. water.
   - add 10 ml 1+1 H$_2$SO$_4$; heat to 95 deg.C. (keep heat above 85
     deg.C through entire procedure)
   - titrate permanganate solution into this till faint pink color
     appears, and persists for 1 minute.
   - calculate:

$$\text{Molarity KMnO4} = \frac{\text{g. Na}_2\text{C}_2\text{O}_4}{\text{ml titrated x .335}}$$

   - adjust to .05M
2. Standardize nitrite stock solution:
   - Add, in this order, 50 ml standard .05 M KMnO$_4$, 5 ml.
     concentrated H$_2$SO$_4$, and 50 ml stock nitrite solution.
   - warm to 80 deg.C on hot plate.
   - Titrate Na$_2$C$_2$O$_4$ into this solution until permanganate color is
     just barely discharged.
   - calculate molarity of stock solution.
3. Prepare working standards (.2-1.0 ppm, or expected range); 50 ml. of
   each.
4. Prepare reduction column:
   - insert glass wool plug into bottom of column; fill with water.
   - add enough Cu-Cd granules to fill to 18.5 cm.
   - wash column with 200 ml dilute NH$_4$Cl-EDTA solution.
   - Activate column by passing through it 100 ml. of a solution
     composed of 25% 1 mg/L NO$_3$-N standard and 75%
     NH$_4$Cl-EDTA solution (7-10 ml/min).
5. Adjust sample pH to between 7 and 9.

6. Reduce sample:
   - to 25 ml. sample add 75 ml. NH_4Cl-EDTA; mix.
   - pour through column (7-10 ml./min); discard first 25 ml.
7. To reduced sample (and nitrite standards) immediately add 2 ml. NED reagent; mix; wait 10 min for color development.
8. Read absorbance on spectrophotometer at 543 nm. Prepare calibration curve and calculate concentration.
9. Prepare nitrate standards and run these through column; test reduced standards for NO_2-N, to verify reduction column efficiency.

Analysis: Nitrate Electrode Method for Nitrate Nitrogen

This electrode has an ion exchange material contained in a solid plastic membrane. It acts as a small ion exchanger which exchanges the nitrate ion wit hydrogen ions across the membrane, creating the voltage potential. A separate reference electrode is needed for this measurement (there will be two probes in the sample solution). The reference electrode is a typical pH reference electrode, but one with a double junction (separate chamber that allows the user to choose a different filling solution).

Quality Control:
- Chloride and bicarbonate ions in substantial amounts interfere with the electrode reading. ISA buffer (Ionic Strength Adjustment Buffer), an acid, is added to the sample to prevent bicarbonate interference (changes bicarbonate to carbonic acid).
- Assure accuracy of standards.
- Test should be done within 48 hours of sampling. The nitrate present is fairly stable, but with standing the ammonia and nitrite in the sample will oxidize, yielding false high results. If septic conditions occur, nitrate will be reduced to nitrogen gas.
- Store Nitrate electrode in 100 ppm Nitrate standard solution. Replace endings monthly; replace electrode after one year.
- Nitrate standards have a shelf life of two weeks.

Apparatus:
- pH meter
- double junction reference electrode
- nitrate selective electrode

Reagents:
nitrate-free distilled water: Pass distilled water through deionizer column appropriate for removing nitrate; store in tightly stoppered glass container. Use for preparing reagents, standards, & for making dilutions.

stock nitrate solution: (the standard)
Dry potassium nitrate (KNO_3) at 105 deg C for 24 hrs. Dissolve .7218 g. in water and dilute to 1 L. (100 mg/L NO_3-N). Prepare fresh every two weeks. Use this solution to prepare working standards.

buffer solution: (prevents bicarbonate interference)
Dissolve 17.32 g. $Al_2(SO_4)_3 \cdot 18H_2O$, 3.43 g. Ag_2SO_4, 1.28 g. H_3BO_3, and 2.52 g. sulfamic acid (H_2NSO_3H) in 800 ml. water. Adjust to pH 3 with .1 N NaOH. Dilute to 1 L and store in dark glass bottle.

sodium hydroxide, .1N: (buffer preparation)
Dissolve 4 g. NaOH in water; dilute to 1 L.

Procedure:
1. Place 20 ml. of standard or sample in a small beaker; add 20 ml buffer. Mix. (Use standards of .1, 1.0, 10, 100 ppm).
2. Immerse electrodes; read millivolts; plot on semilog graph paper.

Chemical Dosage—Jar Testing

Jar testing, a practical trial and error method of determining optimum chemical dosage for solids, BOD and phosphorus removal at wastewater treatment plants, has been the accepted bench testing procedure for many years.

Treatment Plant Significance

Chemical addition at wastewater treatment facilities is performed mainly for the purpose of removing suspended particulate matter from the water. These suspended particles, colloids, are not heavy enough to be settled out, and remain stably in suspension unless biological treatment is employed, or chemical coagulants are added. Chemical coagulation and flocculation are complicated processes, not always easy to optimize, and even more troublesome if the influent water quality is changeable. Effectiveness depends upon type and amount of coagulant, type, length and completeness of mixing and flocculation, temperature, pH, alkalinity of the water, other chemicals present, etc.

Common Metal Coagulants:
Aluminum Sulfate (Alum) $Al_2(SO_4)_3$
Purchased in two forms: dry alum $Al_2(SO_4)_3 \cdot 14H_2O$ (powder or granular)
liquid alum - available in several strengths.

In Reaction: $Al_2(SO_4)_3 \; + \; Ca(HCO_3)_2 \; ---> \; 2Al(OH)_3 \; + \; CaSO_4 \; + \; CO_2$
alum　　　　natural alkalinity　　　floc

Alum dosage is optimum within a pH range of 5.5 - 8.5. The chemical itself is acidic, and has a tendency to lower the pH of the process water.

At wastewater treatment plants, there is usually adequate alkalinity in the water to buffer this added acidity, and coagulant doses of 30-40 ppm are not unusual.

Ferric Chloride, $FeCl_3$

The most frequently used coagulant in wastewater treatment, ferric chloride is available in dry and in liquid form, and the reaction is the same as that for alum; the floc is Ferric Hydroxide $Fe(OH)_3$. Ferric chloride is also very acidic, but works well over a wider pH range than alum does. Its primary disadvantage is its strong yellow color which stains - everything.

Ferrous Chloride, $FeCl_2$

This is a waste product of the steel industry called pickle liquor, and is in some areas a less expensive alternative to ferric chloride. It can be just as effective if it is added at a process point where there is sufficient free oxygen (it oxidizes to ferric chloride). When ferrous chloride is used, it is often added in aeration.

Lime $Ca(OH)_2$

Lime is sometimes used for its coagulant properties, and it acts somewhat differently from the others. It is an alkaline chemical, and reacts with the water to form the fine precipitate $CaCO_3$. This is not a true floc, but has a similar attractive effect on colloidal particles, which group and settle with it. For those waters on which it will work effectively, lime also raises the pH of the water enough to aid in conversion of many inorganic ions (including phosphates) to precipitates, which then settle out. Lime in its dry form is a very fine powder, is only slightly soluble in water, and must be carried as a slurry. Handling is a nuisance.

Various other metal salts may be used for coagulation; some sodium and zinc salts will work effectively, but cost limits use. A disadvantage of all the metal salts used for chemical coagulation is that they are significant additives to the water, and increase the amount of necessary sludge handling.

Polymers

Even at the optimum coagulant concentration, destabilization of the colloidal particles is not always complete, and since the 1950's, addition of a slightly negative polyelectrolyte (polymer) is favored to complete the neutralization. Polymers are water soluble long chain organic molecules.

There are three types of polymers:

Anionic—negatively charged; used with the metal coagulant for removal of TSS, BOD, phosphorus in wastewater; (.2-.5 mg/L).

There are three types of polymers:

Anionic—negatively charged; used with the metal coagulant for removal of TSS, BOD, phosphorus in wastewater; (.2-.5 mg/L).

Cationic—positively charged; used for solids/water separation in sludges (10 lb/dry ton, or 500 mg/L in a 10% solids sludge).

Nonionic—no charge; industrial uses.

Polymers are extremely slippery to the feel. This presents a definite safety hazard when spills occur, either in process or in the laboratory. Most dry polymers have a shelf life of two years. If 100% strength liquid, shelf life is one year. If diluted, shelf life is about 1 week, though this varies depending upon the amount of dilution. Polymers cannot be effectively dissolved in cold weather. The dry particles cling together and form insoluble "fisheyes". Do not attempt to dissolve polymers in hot water. They gel up unevenly.

Analysis: Jar Test Procedure for Chemical Dosage

Jar testing is only an attempt to achieve a ball park approximation of correct chemical dosage for the treatment process. This is done as a batch operation; both mixing and settling are very efficient. Inplant however, with the flow-through mode, inefficient mixing and short circuiting will necessitate adjustments to chemical dosage.

Quality Control:
- With jar testing, it is important to keep environmental conditions as close to those in the treatment plant as possible. Sample water temperature in the jars should be kept the same as process water. Laboratory chemical stock solutions should be prepared with tap water or with the type of water that is used inplant for this purpose.
- Vary dosage of only one coagulant with each jar test run. If polymer is added along with the metal coagulant, hold the polymer dose steady for all jars. Alternately, hold metal coagulant dose steady, and vary polymer dose.

Apparatus:
- jar test apparatus
- 1500 ml jars (beakers)

Reagents:
Coagulant chemical of choice: prepare stock solution of 1000-5000 mg/L, dissolved in tap water.

2. Add increasing amounts of coagulant chemical to each jar, so that the range of additions widely straddles the expected optimal chemical dosage.
3. Rapid mix the jars for about 20 seconds.
4. Set mixers to flocculation speed for about 20 minutes.
5. Stop and withdraw mixers from jars. Allow floc to settle for 30-45 minutes.
6. Decant supernate water from each jar for testing; be careful not to collect any solids that may be lying on water surface.
7. Test water from each jar for parameter of concern to determine ideal chemical dosage.
8. If need is indicated, run the jar test procedure again, choosing chemical dosages that narrow the range.

12

Chlorine Residual

Chlorination is the most widely used means of disinfecting water. When chlorine gas is dissolved into pure water, it forms hypochlorous acid, hypochlorite ion, and hydrogen chloride.

$$Cl_2 + HOH \text{ ------> } HOCl + (OCl)^- + HCl$$

The total concentration of HOCl and $(OCl)^-$ is termed Free Chlorine Residual. This is the disinfectant. These two exist in equilibrium with each other; if the pH of the water is lower, there will be more HOCl. From a disinfection point of view, this is desirable; HOCl is the more powerful disinfectant.

If there is any ammonia in the water the chlorine will react with the ammonia to form chloramines.

$$Cl_2 + NH_3 \text{ ------> } \quad NH_2Cl \quad \text{then} \quad NHCl_2 \quad \text{then} \quad NCl_3$$
$$\text{monochloramine} \qquad \text{dichloramine} \qquad \text{trichloramine}$$

With mole ratios of chlorine:ammonia of 1:1, both monochloramine and dichloramine are formed; concentrations of each are a function of the pH (lower pH, more dichloramine). Trichloramine occurs when greater amounts of chlorine are added, but it is unstable and short-lived.

The total concentration of chloramines is termed Combined Chlorine Residual. It is not nearly as powerful a disinfectant as free chlorine, and a much larger dose and longer contact time is required to provide the same bactericidal action.

Chlorine residual in wastewater effluents is usually all combined chlorine residual.

Treatment Plant Significance

Chlorine discharged by a wastewater treatment plant to the receiving stream may be harmful to stream life, even in small concentrations. Chlorine residual in wastewater effluents is limited. Enough chlorine must be added to meet bacteriologic limits, but excess residual must be dechlorinated.

Treatment Plant Control

Problem: Final effluent Chlorine Residual too low.

Possible Cause	Test This Sample Also For	Other Checks
Final effluent TSS, BOD too high	COD, TSS/VS, NH_3	Raw COD, TSS F/M ratio, MCRT Mixed Liquor TSS TS conc.sludge Sludge blanket level
Incomplete nitrification/ denitrification	NH_3, NO_3, NO_2	Raw COD, TSS/VS, NH_3 TS conc. prim.sludge F/M, MCRT Mixed liquor TSS
Chemical feed problem	Dechlorinating chemical conc.	Flow proportioned control Chlorine diffusers Chlorinator operation Chem. feedwater flow Dechlorinating chem.feed
Incorrect analysis	Use alternate method	QC procedures Refer to Standard Methods

Problem: Final effluent chlorine residual too high.

Possible Cause	Test This Sample Also For	Other Checks
Chemical feed problem	Dechlorinating chemical conc.	Flow proportioned control Chlorinator operation Chemical feedwater flow Dechlorinating chem.feed
Incorrect analysis	Use alternate method	QC procedures Refer to Standard Methods

Analysis: Chlorine Residual
DPD Ferrous Titrimetric Method for Chlorine Residual

When N,N-diethyl-p-phenylenediamine (DPD) is added to a sample containing free chlorine residual, an instantaneous oxidation of the DPD occurs by the chlorine, producing a DPD compound which has a deep pink color. This color can be titrated out with ferrous ion (ferrous ammonium sulfate, FAS), thereby measuring the quantity of Free Chlorine Residual. (ferrous oxidizes to ferric, DPD is reduced back to its original state). A buffer is added to control pH in the 6-7 range where the reaction works best.

If potassium iodide is also added, any chloramines in the sample will oxidize the iodide to free iodine. The iodine will then oxidize the DPD to create more color. Titration with FAS will yield a measurement of Total Chlorine Residual. Subtract the results of the two titrations for a measurement of Combined Chlorine Residual.

Quality Control:
- Add reagents together <u>before</u> adding sample. If not done this way, reagents take 20 minutes or more to react; it will seem like the test isn't working.
- The chemical reaction to liberate free iodine when measuring the combined chlorine segment is slow - let stand 2 minutes before titrating.
- This test is accurate for chlorine residual concentrations up to 5 ppm. If sample chlorine concentration is expected to be greater, dilute the sample.
- If color reverts after titration is completed, ignore it.

Apparatus:
- buret
- erlenmeyer flask

Reagents:
<u>phosphate buffer solution</u>: (pH adjustment)
Dissolve 800 mg disodium EDTA in 100 ml distilled water. Add 24 g. anhydrous Na_2HPO_4 and 46 g. KH_2PO_4 to this. Dilute to 1 L.

<u>DPD indicator solution</u>: (produces the color)
Carefully add 6 ml concentrated H_2SO_4 to 2 ml. distilled water; mix. Add this to 500 ml. distilled water. Add to this 200 mg. disodium EDTA and 1 g. DPD oxalate. Dilute to 1 L. Store in brown bottle. Discard when discolored.

ferrous ammonium sulfate: (the titrant)
Boil 1 L. distilled water for 5 minutes; cool. Bring back to volume. Add
.3 ml. concentrated H_2SO_4 and 1.106 g. $Fe(NH_4)_2(SO_4)_2 \cdot 6H_2O$.   If
standardization is required, refer to Standard Methods, 18th Ed. p.4-44.

potassium iodide: (converts combined chlorine)
Dry crystals

Procedure:

1. Place 5 ml. each of buffer and DPD solution in 250 ml Erlenmeyer
   flask; mix.
2. Add 100 ml. sample; mix.
3. For Free Chlorine Residual, titrate rapidly with FAS titrant until color
   is discharged.
4. For Total Chlorine Residual, omit step #3; add several crystals KI; mix.
   Wait 2 minutes for reaction to proceed; then titrate to discharge color.
5. 1 ml titrant = 1 mg/L chlorine residual
6. Total Chlorine Residual - Free Chlorine Residual = Combined Chlorine
   Residual.

Analysis: Chlorine Residual
DPD Colorimetric Method for Chlorine Residual

The pink color produced by the DPD method can be compared to chlorine
standards and read with a spectrophotometer according to Beer's Law.

Quality Control:
• Assure that standards and samples are treated in exactly the same way.

Apparatus:
• Spectrophotometer (500 nm)

Reagents:
phosphate buffer solution: (pH adjustment)
Dissolve 800 mg disodium EDTA in 100 ml distilled water. Add 24 g.
anhydrous Na_2HPO_4 and 46 g. KH_2PO_4 to this. Dilute to 1 L.

DPD indicator solution: (produces the color)
Carefully add 6 ml concentrated H_2SO_4 to 2 ml. distilled water; mix. Add
this to 500 ml. distilled water. Add to this 200 mg. disodium EDTA and 1
g. DPD oxalate. Dilute to 1 L. Store in brown bottle. Discard when
discolored.

<u>potassium iodide</u>: (converts combined chlorine)
Dry crystals

<u>starch</u>: (titration endpoint indicator)
Add 5 g. starch to 1 L. boiling distilled water. Let stand overnight to settle.
Decant supernate and use.

<u>standard sodium thiosulfate</u>: (titrant, to standardize chlorine solution)
Dissolve 6.205 g. $Na_2S_2O_3 \cdot 5H_2O$ in distilled water. Add .4 g. solid NaOH
and dilute to 1 L. Must be standardized.

<u>chlorine standards</u>: (for colorimetric comparison with sample)
Dilute commercial chlorine to approximately 100 ppm.

Procedure:
1. Standardize chlorine solution:
   - Add 2 ml. acetic acid and 25 ml. distilled water to a flask.
   - Add 1 g. KI and 100 ml chlorine solution.
   - Titrate with .025 N $Na_2S_2O_3$ till yellow color is almost gone.
   - Add 1 ml. starch; continue titrating till blue color is discharged.
   - Run a blank through the titration also.
   - *Calculate concentration of standard solution:*

$$\text{mg/L } Cl_2 = \frac{(\text{ml.titr. } Cl_2 \text{ sol'n - ml titr. blk.}) \times N \text{ thiosulfate} \times 35450}{\text{ml. sample}}$$

2. Prepare working standards to cover range.
3. Set up flasks for each standard, reagent blank, samples.
   Add 5 ml. DPD and 5 ml. buffer to each; mix.
4. Add standards and samples to flasks; mix.
5. If Total Chlorine Residual is desired, also add a few crystals of KI.
   Wait two minutes for color development.
6. Read absorbance on spectrophotometer. Prepare calibration curve and
   calculate sample concentration.

Analysis: Chlorine Residual
Amperometric Titration Method for Chlorine Residual

An amperometric titrator consists of a pair of electrodes connected by a salt
bridge. One electrode senses the concentration of ionized chlorine in the solution,
and becomes polarized by it. A current flow then occurs to the opposite electrode,
which is picked up by a microammeter, producing a needle reading. The intensity

of this reading decreases steadily (needle drops) as the ionized chlorine is neutralized by the reagent, phenylarsine oxide. When the needle stops moving, the current has stopped, and the chlorine has all been neutralized. At that point the titration is ended, and the ml. used is read.

The change in amperage is observed as free chlorine is reduced by the titrant phenylarsine oxide. The reaction is sluggish over pH 7.5, and a pH 7 buffer is added to control pH and convert all free chlorine to OCl ion. When the amperage stops changing, all of the free chlorine has been reduced, and the titration is stopped.

At pH levels below 6, chloramines are reduced, and can be measured indirectly (iodometrically). If KI is added, the chloramines oxidize iodide to free iodine. The phenylarsine oxide titrant then reduces the free iodine, thereby measuring the amount of chloramines present. The pH is controlled for this titration with a pH 4 buffer. By conducting a two-stage titration, first at pH 7, then with KI at pH 4, free and combined chlorine can be separated.

Amperometric titration is not subject to interference from color or turbidity.

Quality Control:
- This method measures accurately down to .01 ppm chlorine residual. To measure smaller concentrations, refer to Standard Methods, 18th Ed. 4(44-45).
- Amperometric titrator electrodes may need periodic cleaning with an abrasive non-chlorine cleanser. See manual provided with electrodes.
- Buffers used in this procedure must be recently prepared. If bacterial growth has occurred in the buffers, a chlorine demand will be established which will lead to inaccurate results.
- Phenylarsine is toxic. Handle carefully.

Apparatus:
- amperometric titrator.

Reagents:
phenylarsine oxide, .00564N: (the titrant)
This reagent is best purchased from a chemical distributor as a solution. Preparation is tedious, and safe handling of the highly toxic powder requires extreme care.

pH 4 buffer: (to titrate combined chlorine residual).

pH 7 buffer: (to titrate free chlorine residual).

potassium iodide solution: (converts to free iodine for combined chlorine titration)
Dissolve 50 g. KI in distilled water; dilute to 1 L. Store in dark bottle.

Procedure:
1. For free chlorine residual: To 200 ml sample, add 1 ml pH 7 buffer. Titrate with phenylarsine oxide in progressively decreasing volumes, observing current change on titrator. Stop titrating when current stops changing. Record ml. titrated.
2. For combined chlorine residual: To remaining sample, add 1 ml. KI and 1 ml. pH 4 buffer. Continue titrating to endpoint, as above.
3. Total ml. titrated = Total Chlorine Residual
   Total ml. titrated - ml. titrated #1 = Combined Chlorine Residual

Analysis: Chlorine Residual
Ion Specific Electrode Method for Chlorine

Measurement by ion specific electrode is an indirect (iodometric) measurement of chlorine. The electrode develops a voltage which is dependent upon the difference in concentrations of iodine and iodide in the solution. KI and acid buffer are added to the sample. Chlorine in the sample changes the KI to free iodine. KIO_3 is used as a standard in place of chlorine; it is converted to iodine, which is measured.

Quality Control:
- Add reagents, then wait 5 minutes !
- Mix standard, acid, KI together first; add these to sample; this shortens the waiting time.
- The stock KI solution should be colorless; if it turns yellow at all, discard.

Apparatus:
- pH/millivolt meter
- combination reference/iodine selective electrode (chlorine residual)

Reagents:
pH 4 buffer: (to titrate combined chlorine residual).

pH 7 buffer: (to titrate free chlorine residual).

potassium iodide solution: (converts to free iodine for combined chlorine titration)
Dissolve 42 g. KI and .2 g. Na_2CO_3 in 500 ml distilled water. Store in dark bottle.

potassium iodate solution, .00281N: (the standard)
Dissolve .1002 g. KIO_3 in distilled water; dilute to 1L.

Procedure:
1. Pipet .2, 1.0, 5.0 ml iodate solution into 100 ml stoppered volumetric flasks
2. Add to each flask 1 ml. pH4 buffer and 1 ml. KI solution. Mix. Let stand 2 minutes
3. Dilute each to 100 ml. with distilled water; mix. Pour into beakers, insert electrode into first standard (.2 ml, .2 mg/L); record millivolts.
4. Rinse electrode; repeat for each standard, and samples.
5. Prepare calibration curve on semilog graph paper.

13

Volatile Acids/Alkalinity

Used as an indicator of potential upset in the anaerobic digester, the Volatile Acids/Alkalinity test is performed on primary digester liquid (supernatant).

Volatile Acids—short chain organic acids which are intermediate products of the anaerobic breakdown of carbohydrates, proteins, fats. Acetic, propionic, butyric acids are typical; they can be measured by neutralization with a basic titrant. An overabundance of these acids can have an inhibiting effect on anaerobic digestion, if the digester's buffering capacity is exceeded.

Alkalinity—a water's capacity to resist a change in pH when acid is added. In digestion, it is the sludge's capacity to neutralize volatile acids as they are formed. In the laboratory, alkalinity can be measured by neutralizing with an acid titrant to a pH of 4.5, below which no alkalinity exists.

Treatment Plant Significance

Large numbers of saprophytic anaerobes (acid forming bacteria) enter the treatment process with the raw sewage, and are settled out with the primary sludge. These are hardy bacteria, which start decomposing the solids as they lay in the bottom of the primary clarifier. As the sludge enters the digester, these bacteria are already thriving, decomposing organics to volatile acids.

Methane forming bacteria, also strict anaerobes, enter with the raw wastewater, and settle with the sludge. These bacteria, however, are very sensitive to environmental conditions, and remain dormant, until ideal growth conditions take place. It is control of anaerobic digestion which provides these ideal conditions.

Low and constant sludge loading, constant temperature in the mesophilic range, and efficient mixing promotes activity of the methane formers. Acids must not be allowed to accumulate. As the pH drops, all bacteria are inhibited, gas production decreases, and gas concentration becomes more carbon dioxide than methane There must be a population balance in the digester. The Volatile Acids/Alkalinity test results state what it is - as a chemical ratio. Volatile Acid/Alkalinity Ratio is the first measurable change that takes place in an anaerobic digester when it is becoming upset.

Expected Volatile Acids/Alkalinity Ratio in Anaerobic Digester: 1/10 (.1)

Treatment Plant Control

Problem: Higher than normal Volatile Acids/Alkalinity Ratio in anaerobically digesting sludge.

Possible Cause	Test This Sample Also For	Other Checks
Organic Overload	TS/VS	COD,TSS raw wastewater, prim.eff. Prim.sludge TS/VS Digester gas production Gas %CO_2 Supernatant COD, TSS
Primary sludge flow too high, or sludge solids too dilute	TS/VS	Plant flow Prim.sludge pumping frequency Prim.sludge TS/VS Gas production, %CO_2 WAS to digester?
Digester temperature change	TS/VS	Primary digester temperature Gas production, %CO_2 Supernatant COD, TSS
Poor digester mixing	TS/VS	Operation of digester mixing equipment Gas production, %CO_2 Supernatant COD, TSS Scum blanket depth
Incorrect analysis		QC procedures Refer to Standard Methods

With anaerobic digester control, all tests and parameters should be checked for the past 30 days.

Analysis: Titration Method for Determination of Volatile Acids/Alkalinity

Volatile Acids/Alkalinity test is performed weekly for routine anaerobic digester control (more often if digester is upset). In this test, volatile acids and total alkalinity are separately measured and then set together as a Volatile Acids/ Alkalinity ratio. An efficiently operating anaerobic digester should have a VA/A ratio of about 1:10 (also written as .1) The value of the test results is in recognition of a trend toward upset. A consistently larger concentration of alkalinity ensures that there is adequate buffering capacity in the sludge.

Quality Control:
- This test is based on pH readings. No chemicals should be added to the sample to aid separation; all coagulants and most polymers will change the initial pH, and lead to inaccurate results.
- Calibrate meter before each use.
- Standardize acid and base titrants.
- Titrate carefully. It is easy to overrun these titrations. There are inflection points at which the pH changes rapidly; they mark a change in the type of alkalinity being measured.

Apparatus:
- pH meter; pH probe
- magnetic stirrer
- thermometer
- hot plate

Reagents:
pH 7 buffer solution: (meter calibration).

pH 4 buffer solution: (meter calibration).

standard sulfuric acid, .05 N: (titrant; to measure alkalinity)
Add 7 ml. concentrated H_2SO_4 to 500 ml. distilled water; dilute to 1 L. Mix well. Take 200 ml of this and make up to 1 L. Must be standardized.

sodium carbonate, .05 N: (to standardize the H_2SO_4)
Dissolve 2.65 g. anhydrous Na_2CO_3, dried at 140 deg.C into 500 ml distilled water; dilute to 1 L.

standard sodium hydroxide, .05 N: (titrant; to measure volatile acids)
Dissolve 2 g. NaOH in distilled water; dilute to 1 L.

phenolphthalein indicator solution: (to detect volatile acid endpoint)
Dissolve 80 mg. phenolphthalein in 100 ml. absolute methanol.

methyl orange indicator solution: (to detect alkalinity endpoint)
Dissolve 500 mg. methyl orange in distilled water; dilute to 1 L.

Procedure:
1. Standardize the sulfuric acid:
   - To 25 ml .05N Na_2CO_3 add three drops methyl orange and titrate with .05 N H_2SO_4 to the color change.
   - calculate normality H_2SO_4. Adjust to .05 N.
2. Standardize the sodium hydroxide:
   - To 25 ml. .05 N H_2SO_4 add three drops phenolphthalein and titrate with .05 N NaOH to the appearance of faint pink color.
   - calculate normality of NaOH. Adjust to .05 N.
3. Sample of digesting sludge must be separated from its water. Let it stand till separation is achieved, or centrifuge to separate.
4. Calibrate pH meter.
5. Take 50 ml of supernate water and read pH.
6. Titrate with .05 N H_2SO_4 to pH 4.0. Record ml.used.
7. Continue titration to pH 3.3.
8. Lightly boil sample for 5 minutes under watch glass to drive off CO_2. Cool to original temperature.
9. Titrate sample back to pH 4 with .05N NaOH and mark buret reading.
10. Titrate sample up to original pH with .05N NaOH; record ml. used between pH 4 and end of titration.
11. *Calculate:*

    Total Alkalinity(mg/L) = ml. H_2SO_4 titrated x 50

    Volatile Acid Alkalinity(mg/L) = ml NaOH titrated x 50

    If Volatile Acid Alkalinity is > 180, multiply the value by 1.5 for the true Volatile Acids value.

    If Volatile Acid Alkalinity is < 180, then the Volatile Acid Alkalinity is the true Volatile Acids value.

    $$\frac{\text{Volatile Acids, mg/L}}{\text{Total Alkalinity, mg/L}} = \text{Vol.Acid/Alk. Ratio}$$

Appendix: Practice Problems

General

1. An empty container weighs 2 kg. Filled with water it weighs 23 lb. What is the volume of the container (liters)?

    Answer: 8.44 liters

2. Change to concentration in mg/L:
    a. .2 lb/L
    b. 1 lb/gal
    c. .006%
    d. 12 g/MG

    Answer: a. 90,800 mg/L
    b. 119,947 mg/L
    c. 60 mg/L
    d. .003 mg/L

3. Change 65 ppm to units of % concentration.

    Answer: .0065%

4. One gram of salt is dissolved in 30 gallons of water. What is the concentration in ppm?

    Answer: 8.8 mg/L

5. How many grams is 3.5 lb. equivalent to?

   Answer: 1589 grams

6. What is the normality of a 500 mg/L solution of phosphoric acid?

   Answer: .015 N

7. A solution is prepared at a concentration of 400 μg/L.
   What is the concentration in mg/L?

   Answer: .4 mg/L

8. A .06 N solution of sulfuric acid is what concentration in mg/L ?

   Answer: 2940 mg/L

9. A .05 N solution of sulfuric acid has been prepared. Find:
   a. the molarity
   b. the concentration in mg/L
   c. the % concentration
   d. how much of it is needed to prepare 50 ml of a .012 N solution

   Answer: a. .025 M
             b. 2450 mg/L
             c. .2450%
             d. 12 ml.

10. How much of a .5 N sulfuric acid solution is needed to prepare 100 ml of a
    .125 N solution?

    Answer: 25 ml

11. What is the mg/L concentration of a 1 N HCl solution?

    Answer: 36,000 mg/L

12. What is the molarity of a 450 ml HCl solution which contains 20 grams of HCl?

    Answer: 1.24 M

13. How many ml. of .3 N HCl will neutralize 50 ml of .0125 N NaOH?

    Answer: 2.1 ml.

14. Twenty ml. of 2N HCl is mixed with 480 ml. of .5N HCl.
    What is the concentration of the final mixture?

    Answer: .56N

15. If 50 ml. of a .5N NaOH solution are required to neutralize 25 ml. of acid solution, what is the normality of the acid ?

    Answer: 1N

16. How many ml. of water are needed to reduce 500 ml of a sodium sulfate solution (1000 mg/L) to:
    a. 80 ppm
    b. .01N

    Answer: a. 5750 ml.
             b. 200 ml.

17. What weight of dibasic potassium phosphate is needed to prepare one liter of a 1000 mg/l solution of potassium?

    Answer: 2.22 grams

18. Concentrated hydrochloric acid has a specific gravity of 1.18. The solution, as purchased, is 37% acid. How many ml. of this acid are needed to prepare 500 ml of a .2N HCl solution?

    Answer: 8.24 ml.

19. Twenty ml. of element X (approximately .01N) should neutralize 20 ml. of element Y (exactly .01N) in a titration. However, it actually takes 26 ml. of element X to complete the neutralization. What is the normality of element X?

Answer: .008N

20. Thirty eight mg. of chemical is dissolved in 30 ml water. What is the concentration in mg/L?

Answer: 1267 mg/L

Solids

1. Calculate mg/L total suspended solids and % Volatile in the following samples. List what type of wastewater sample each is most likely to be.

   a. tare wt.   23.0788 g.
      dry wt.   23.0851 g.
      ash wt.   23.0811 g.
      sample size 50 ml.

   b. tare wt.   23.9123 g.
      dry wt.   23.9138 g.
      ash wt.   23.9127 g.
      sample size 100 ml.

   c. tare wt.   22.1798 g.
      dry wt.   22.1874 g.
      ash wt.   22.1824 g.
      sample size 30 ml.

   Answer: a. TSS 126 mg/L
                  64 % Vol.
                  primary effluent
               b. TSS 15 mg/L
                  73 % Vol.
                  final effluent
               c. TSS 253 mg/L
                  66 % Vol.
                  raw influent

2. Referring to sample a. from the previous question, if this were the influent sample at a 30 MGD treatment plant, and it was required to remove 85% of the total suspended solids in treatment, how many pounds of total suspended solids would it be necessary to remove in a month's worth of treatment?

   Answer: 803,890 lb.

3. Total solids analysis of a sludge yielded this data:

tare wt.	40 g.
wet wt.	270 g.
dry wt.	54 g.
ignited wt.	45 g.

   Calculate the % total solids in the sludge, and the % volatile.

   Answer: 6.1% TS
             64% VS

4. A 100 ml sample of sludge gave the following results on total solids analysis:

tare wt.	42.0 g.
wet wt.	142.0 g.
dry wt.	52.0 g.
ignited wt.	44.0 g.

   If 1000 pounds of this sludge were incinerated, how many pounds of ash would have to be disposed of?

   Answer: 20 lb.

5. A domestic wastewater contains 350 mg/L total suspended solids. Primary sedimentation facilities remove 65% of these solids. How many gallons of primary sludge containing 5% solids will be produced per million gallons of wastewater settled?

   Answer: 4550 gal.

6. Operating with a flow of 5 MGD, an influent total solids content of 10(mg/L, a total dissolved solids content of 650 mg/L, and a 10 ppm NPDF permit limit for total suspended solids in the final effluent, what percentage total suspended solids would be removed from the water before discharge?

Answer: 97%

7. The raw sewage entering the .5 MGD treatment plant has a total suspend solids concentration of 200 ppm. The primary sludge pump discharges gpm for 10 minutes - 4 times/day. Primary sludge is 3% solids. If operatioı are normal, what is the TSS concentration of the primary effluent?

Answer: 80 mg/L

8. Treatment plant flow is 4 MGD. Primary effluent total suspended solids 100 mg/l. Mixed liquor total suspended solids is 2000 mg/l and is 70 volatile, in a 1 MGD aeration tank.
   a. How many lb. MLSS are under aeration?
   b. How many lb. MLVSS are under aeration?
   c. How many lb. TSS are fed to aeration each day?

Answer: a. 16680 lb.
             b. 11676 lb.
             c. 3336 lb.

9. On a 30 minute settling test, the activated sludge settled to 350 ml. in a oı liter graduate. MLSS concentration is 3000 mg/L. Plant flow is 2 MGD.
   a. What is the SVI?
   b. What is the SDI?
   c. What is the activated sludge volume produced/day?
   d. What is the concentration of the activated sludge?

Answer: a. SVI 117
             b. SDI .86
             c. 700,000gal
             d. .86%

10. Each day 2800 gal of raw sludge is pumped into the anaerobic digester. This influent is 6.5% total solids, 68% volatile. The digested sludge is 9.5% total solids, 54% volatile.

    Calculate:
    a. pounds total solids raw sludge pumped/day
    b. pounds volatile solids raw sludge pumped/day
    c. gallons digested sludge produced/day
    d. gallons supernatant produced/day
    e. pounds total solids digested sludge produced/day

    Answer: a. 1518 lb.
                   b. 1032 lb.
                   c. 1326 gal
                   d. 1474 gal
                   e. 1051 lb.

Biochemical Oxygen Demand

1. In preparing a BOD test, twenty ml. of sample was added to 280 ml. dilution water in the BOD bottle. What was the percent dilution?

    Answer: .067

2. BOD test results yield:

    Primary Effluent:
    Initial DO   8.00 mg/L
    Final DO   3.00 mg/L
    ml. sample   10

    Final Effluent:
    Initial DO   8.00 mg/L
    Final DO   5.00 mg/L
    ml. sample   40, plus 3 ml. seed (primary effluent)

    What is the BOD of the Final Effluent?

    Answer: 10.5 mg/L

3. A 3% sample of Primary Effluent was found to contain a BOD of 110 mg/L
Three ml. of the primary effluent sample was used to seed a final effluer
sample. What was the loss due to seed?

Answer: 1.1 mg/L

4. An industrial waste was analyzed for BOD. Initial DO was 8.2 mg/L; Fina
DO was 6.5 mg/L. Dilution was .01%.
   a. Calculate BOD
   b. How many pounds of BOD are contained in 1500 gallons of this waste?

Answer: a. 17,000 mg/L
           b. 213 lb.

5. BOD is being tested on a new industrial waste sample; it is expected to b
about 2000 mg/L. In performing this test, how many ml. of sample shoul
be put in the BOD bottle?

Answer: .3 ml-1.2 ml.

6. Calculate the BOD of this Raw wastewater.

   DO in blank after incubation 8.3 mg/L
   DO in dilution after incubation 3.7 mg/L
   DO in sample before dilution 8.5 mg/L
   % sample in dilution 2.5%

Answer: 184 mg/L

7. A BOD on a 90 ml. sample of final effluent was seeded with 4 ml. of wate
(BOD=250 ppm). If the total DO depletion in the final effluent bottle was 6.
ppm, and the DO depletion for each ml. of seed was .2 ppm, what was th
BOD of the final effluent?

Answer: 17.2 mg/L

8. A clean river sample is to be analyzed for BOD. Expected BOD concentration is 5 ppm. Water temperature is 8 degrees C. DO content is 12 ppm. pH is 7.0.
   a. What should be done to prepare this sample for the BOD test?
   b. How much should it be diluted for the BOD test?

   Answer: a. Warm to 20C.
               Shake to liberate supersat. DO.
               Seed sample.
           b. Don't dilute at all.

9. River flows at 3.1 cubic feet per second; it has a natural BOD concentration of 5 mg/L upstream from the discharges of a municipal wastewater treatment plant and an industrial wastewater treatment plant. The municipal discharge is a .5 MGD flow with a BOD concentration of 20 mg/L. The industry discharges a .8 MGD flow with a BOD concentration of 60 mg/L. What is the BOD concentration in the river downstream from these discharges?

   Answer: 20.6 mg/L

10. A raw wastewater sample was tested for BOD. Oxygen depletion on the 6 ml sample was 4.5 mg/L. If 30% of this BOD was settled out in the primary clarifier, what is the BOD of the primary effluent?

    Answer: 135 mg/L

Phosphorus

1. In preparing for a jar test, how many ml. of a 2% coagulant stock solution should be added to a 1000 ml. jar to achieve a concentration of 25 mg/L?

   Answer: 1.25 ml.

2. From a 500 mg/L stock phosphate solution, how many ml. would be needed to prepare 100 ml. of a working standard of 2.5 ppm?

   Answer: .5 ml.

3. A colorimetric procedure for phosphorus determination is graphed. Standard slope is 264. Absorbance of sample reads .023. What is the concentration of phosphorus in the sample?

   Answer: 6 mg/L

4. To prepare one liter of a solution containing 1000 mg/L phosphate for a standard, dry $NaH_2PO_4 \cdot 5H_2O$ is used. How much of this chemical is needed?

   Answer: 2.212 g.

5. By jar testing it was determined that 3 ml. of a 19 g/L alum solution in a 1000 ml. sample would be the correct dosage for phosphorus removal. How many pounds per day of the dry chemical must be added to a 700 gal/min. flow to simulate this jar test dosage?

   Answer: 475.4 lb/day

6. The city has a population of 50,000. If the wastewater treatment plant discharge permit mandates phosphorus removal down to 1 ppm in the discharge water, and the phosphorus content of the influent is 6 ppm, how many pounds of phosphorus are being removed from the water each day? (assume 100 gal/day/capita)

   Answer: 208.5 lb/day

7. If .2 lb. of coagulant is mixed with each gallon of water to prepare a stock solution for feeding into the plant flow, what is the percent concentration of the coagulant in the stock solution?

   Answer: 23989 mg/L

8. For a jar test, .05 lb. coagulant chemical is dissolved in a liter of water. Three ml. of this stock is added to a one liter jar for flocculation. What is the concentration of the coagulant in the jar?

   Answer: 68 mg/L

9. An industrial wastewater treatment plant has a raw wastewater total phosphorus content of 8 ppm. If this plant has an NPDES discharge limit on total phosphorus of 1 ppm, what % removal of total phosphorus must it achieve?

   Answer: 87.5%

10. Total phosphorus concentration of the raw wastewater of an 8 MGD flow is 5.3 ppm. Eighty two percent of this is orthophosphate. How many lb/day of combined organic and polyphosphate are entering the treatment plant?

    Answer: 63.7 lb./day

Nitrogen

1. Directions to prepare an ammonia stock solution specify dissolving 11.457 grams of dried NH_4Cl in a liter of water. Then 10 ml of this stock solution is diluted to one liter with water to make the standard.
   a. What is the concentration of nitrogen in the stock solution ?
   b. What is the concentration of nitrogen in the standard ?
   c. What is the concentration of ammonia in the stock solution ?
   d. What is the concentration of ammonia in the standard ?

   Answer: a. 3025 ppm
              b. 30.25 ppm
              c. 3895 ppm
              d. 38.95 ppm

2. In a colorimetric test for ammonia nitrogen, the sample absorbance was .18; the standard slope calculated out to 24, and the sample has been diluted four times with distilled water. What is the ammonia concentration in the original sample?

   Answer: 17.28 mg/L

3. To prepare 250 ml. of a potassium nitrate solution which is 100 mg/L nitrogen, how much of the chemical should be weighed out?

   Answer: 180.4 mg

4. How many ml. of a .2% solution of potassium nitrate are needed to prepare 100 ml. of a 50 ppm solution of potassium nitrate?

   Answer: 2.5 ml.

5. What is the mg/L concentration of a 1.3 N solution of ammonium hydroxide?

   Answer: 45,500 mg/L

6. Two final effluent samples are tested for nitrate. The following data is obtained:

   Standards (ppm concentration)   .5   .7   .9
   Standards (absorbance reading)   .72   1.2   1.65
   Sample #1 (absorbance)   2.84
   Sample #2 (absorbance)   .98

   Part of this test should be done over. What is wrong here? What should be done to correct it?

   Answer: Sample #1 absorbance is out of range of the standards; the sample should be diluted to read within the standards.

7. One liter of ammonium chloride solution has been prepared; its concentration is 250 ppm NH_3-N. How many mg. of ammonium chloride has been dissolved in the liter of water?

   Answer: 947 mg.

8. A colorimetric procedure reads 500 micrograms of ammonia in the 50 ml sample. What is the ammonia concentration in mg/L?

   Answer: 10 mg/L

9. The ammonia concentration in the raw wastewater is 32 ppm. If the plant flow is 1 MGD, and 220 pounds of ammonia are removed in treatment, what is the ammonia concentration in the final effluent?

   Answer: 5.6 mg/L

10. The potable water treatment plant is drawing from a river with a natural flow of 30 cfs. The normal nitrate content of the river water is 2 ppm. Upstream, the 15 MGD wastewater treatment plant is discharging an effluent with a nitrate concentration of 25 ppm. What is the nitrate concentration of the water which the low lift pumps are bringing into the potable water treatment plant?

Answer: 12 ppm

Chlorine Residual

1. An industrial liquid chlorine is 12% hypochlorite ion (active chlorine). What is this concentration in mg/L?

Answer: 12,000 mg/L

2. A 2% sodium hypochlorite solution is used to prepare a standard for chlorine residual analysis. How many ml is needed to prepare a 100 ppm concentration in a 500 ml flask?

Answer: 2.5 ml

3. What is the percentage of active chlorine in a tank of chlorine gas?

Answer: 100%

4. What is the chlorine demand of a water whose chlorine residual is .5 ppm, if a chlorine dose of 14 lb/day is being added to the 500 gpm flow?

Answer: 1.8 ppm

5. Twenty ml of a 100 ppm chlorine standard is added to a liter of wastewater. After the appropriate contact time, the chlorine residual is read at .7 ppm. What was the chlorine demand of this water?

Answer: 1.3 ppm

6. How many ml. of sodium hypochlorite (12% available chlorine) must be adde to a liter of water to prepare a 1000 ppm available chlorine solution?

   Answer: 8.3 ml.

7. If 15 lb. of chlorine gas are dissolved into 1500 gallons of pure water, wh is the chlorine concentration in ppm?

   Answer: 1199 ppm

8. How many lb. of free available chlorine are there in 20 lb. of $Ca(OCl)_2$?

   Answer: 14.2 lb.

9. How many grams of calcium hypochlorite must be dissolved in a liter of wat to prepare a .1% available chlorine solution?

   Answer: 1.4 grams

10. The chlorine demand of a wastewater is 8 ppm. How many gallons of 15' sodium hypochlorite weighing 9.3 lb/gal must be added each day to the MGD flow to achieve a chlorine residual of 1 ppm?

    Answer: 323 gal.

Appendix: Useful Conversions

Weight:	454 grams	= 1 pound
	1000 μg	= 1 mg.
	1000 mg	= 1 gram
	1000 grams	= 1 kg.
Volume:	1000 ml	= 1 liter
	3785 ml	= 1 gallon
	7.48 cu.ft.	= 1 gallon
Length:	1000 nm	= 1 um
	1000 um	= 1 mm
	1000 mm	= 1 meter
Temperature:	20° Celsius	= 68° Fahrenheit
	0° Celsius	= 32° Fahrenheit
	100° Celsius	= 212° Fahrenheit
If It's Water:	8.34 pounds	= 1 gallon
	1 ml.	= 1 gram
	1,000,000 mg	= 1 liter

Appendix: Bibliography

Adams, V. Water and Wastewater Examination Manual. Chelsea: Lewis Publishers, 1990.

American Water Works Association. Basic Science Concepts and Applications. Denver: American Water Works Association, 1980.

American Public Health Association, American Water Works Association, Water Environment Federation. Standard Methods For the Examination of Water and Wastewater, 18th Edition. Washington D.C.: American Public Health Association, American Water Works Association, Water Environment Federation, 1992.

American Public Health Association, American Water Works Association, Water Pollution Control Federation. Selected Chemical and Physical Standard Methods for Students. Washington D.C.: American Public health Association, American Water Works Association, Water Pollution Control Federation, 1986.

Environmental Protection Agency. Operation of Wastewater Treatment Plants, Vol.I,II. Sacramento: California State University, 1989.

Environmental Protection Agency. Advanced Wastewater Treatment. Sacramento: California State University, 1995.

Hach Co. Water Analysis Handbook. Loveland: Hach Co., 1992.

Hajian, Sr., H. and Pecsok, R. Modern Chemical Technology, Vol.I. Englewood Cliffs: Prentice-Hall, 1988.

Kenkel, J. Analytical Chemistry For Technicians, Second Edition. Boca Raton: Lewis Publishers, 1994.

Jackson, G. Water Chemistry Manual for Water and Spentwater Personnel. Oklahoma City: The Chemists Group, Inc., 1990.

Manahan, S. Environmental Chemistry, 4th Edition. Chelsea: Lewis Publishers, 1990.

Masterton, W., Slowinski, and Stanitski, C. Chemical Principles. Philadelphia: Saunders College Publishing, 1983.

Merck & Co., Inc. Merck Index, 11th Edition. Rahway: Merck & Co., Inc., 1989.

Occupational Safety and Health Administration. Hazard Communication, 29CFR 1910.1200: U.S. Department of Health, Department of Labor, 1970.

Sawyer, C. and McCarty, P. Chemistry For Environmental Engineering. New York: McGraw-Hill, 1978.

Stein, R. Laboratory Manual for Water Pollution. (Unpublished).

Water Pollution Control Federation. Simplified Laboratory Procedures for Wastewater Examination, Third Edition. Washington D.C.: Water Pollution Control Federation, 1985.

Index